CAREER
SUCCESS
IN ENGINEERING

A Guide for Students and New Professionals

BERNARD R. BERSON, PE, LS, FNSPE and
DOUGLAS E. BENNER, PE, FNSPE

KAPLAN) AEC EDUCATION

President, Kaplan Publishing: Roy Lipner
Vice President and Publisher: Maureen McMahon
Acquisitions Editor: Victoria Smith
Production Editor: Karen Goodfriend
Typesetter: Todd Bowman
Cover Designer: Sue Giroux, SLG Creative

Published by Kaplan Publishing,
a division of Kaplan, Inc.

Printed in the United States of America
07 08 09 10 9 8 7 6 5 4 3 2 1

Library of Congress Cataloging-in-Publication Data
Benner, Douglas E.
    Career success in engineering : a guide for students and new professionals / Douglas E. Benner, Bernard R. Berson. — 1st ed.
      p. cm.
    ISBN-13: 978-1-4195-8439-8
    ISBN-10: 1-4195-8439-1
    1. Engineering—Vocational guidance. I. Berson, Bernard R. II. Title.
TA157.B447 2007
620.0023—dc22

                                        2006027368

Kaplan Publishing books are available at special quantity discounts to use for sales promotions, employee premiums, or educational purposes. Please call our Special Sales Department to order or for more information at 800-621-9621, ext. 4444, e-mail *kaplanpubsales@kaplan.com*, or write to Kaplan Publishing, 30 South Wacker Drive, Suite 2500, Chicago, IL 60606-7481.

C o n t e n t s

PART ONE
GETTING STARTED

PART TWO
ON THE JOB

Like most young engineers, when I graduated from college, I had a very limited knowledge about what to expect in working as an engineer. I was the only college graduate from a blue-collar family, so my parents could offer me little information. My only prior experience was one year as a field service technician and one summer as a student intern in the utility industry. This experience was helpful, but in retrospect it did not provide much insight into what engineering was all about.

Having received very little coaching by my professors in what to expect in an engineering position, I assumed that engineering work would be more or less a continuation of college-level engineering lab reports and design projects.

When I started working in industry, I was fortunate to be a part of an extended, formalized training program. Here I learned about the industry, about the technologies involved, about business economics, and about some soft skills such as public speaking and basic "people skills." I was fortunate because such programs are very rare in today's strongly competitive business environment. After completing this program, I received my assignment and jumped in feet first.

There was no formal mentoring program and there was very little guidance in dealing with office and corporate politics. No one provided instructions on how to plan and develop a career. There were no self-help books for young engineers. I was left to my own ingenuity to discover these secrets. Admittedly, I floundered for a few years.

This book is a guide and a tool for the engineering student and the recently graduated engineer to help them hit the deck running in the early phases of their career. We are providing an up-to-date reference manual that deals with the conventional wisdom of engineering careers as well as the rapid changes in today's economy and engineering world.

Keep in mind that all your career development efforts may have to be modified or restarted when you change jobs. And job change is more frequent and more acceptable today than it was 25 years ago.

Many technical as well as nontechnical skills must be assimilated along the way in your engineering career. The appropriate mix of these skills is highly dependent upon your chosen economic sector, practice area, and technical specialty. Today's effective engineer needs to be technically competent and current, must learn to work in a multicultural and multinational environment, must work with multidisciplinary projects, must understand business needs and economics, must learn and work with the organization's culture, and must maintain integrity in engineering ethics.

In this book we provide a somewhat broad-brush review of many requisite career development tools. We also provide some direction regarding how to research and select the best development programs.

Remember, when your parents started their working careers, their employers or unions frequently provided clear career plans. Perhaps they went so far as to systematically analyze an employee's unique development needs. Today, career planning and development is *your* job. This book is intended to help you do that job more effectively.

Douglas E. Benner, PE, F.NSPE

Bernard Berson

I would like to thank Victoria Smith, Acquisitions Editor for Publishing, who invited me to participate in authoring this book, who provided preliminary editing and constructive criticism, and who acted as a cheerleader throughout the sometimes difficult task of continuing production of chapters.

I also want to thank the following people:

Doug Benner, my co-author, whose collaborative effort and organizational skills were immeasurably valuable.

Ira Dorman, Esq., patent attorney, who reviewed my work on patents, copyrights, trademarks, and trade secrets, and who gave valuable input and guidance on that topic.

Arthur Schwartz, Esq., general counsel and deputy executive director of NSPE, who furnished valuable guidance in searching for reference materials on many of the topics in the book.

Michael F. Dolan, Esq., partner in Hoagland, Longo, Moran, Dunst and Doukas, for reviewing the chapters dealing with professional liability and risk management.

My wife, Toby Berson, who was forced to give up a lot of my time and attention during the year I spent writing this book.

I would also like to acknowledge the multitude of people who have mentored me throughout my career and who contributed to my professional growth, providing me with much of the background that enabled me to take on this project.

I want to offer a special thank you to John Gregorits, P.E., who passed away on November 22, 2002. Mentioned in Chapter 20 on mentoring, he was the gentleman who greeted me at my first NJSPE chapter meeting about 40 years ago. His dedication to and zealous efforts for the profession of engineering have been major motivating forces for me. If I had not met John, I probably would not have done the things that resulted in me having the opportunity to write my part of this book. His values live on through me and through the many others he touched during his life.

Doug Benner

I would like to dedicate this work to my wife, Fabiola, who provided unlimited encouragement, understanding, and patience during the writing process. I thank my entire family for their patience and understanding.

I would also like to thank the following individuals:

Bernie Berson, for his leadership, inspiration, and collaboration in producing this work.

My daughter, Laura Benner, for her assistance with research on human resources issues.

My friend and former coworker Harry T. Roman for his assistance and mentoring in the book-writing process.

Dennis Alderfer, CPA, for his assistance and review of the business and accounting issues.

Terry Foster, PhD, of the University of Nebraska–Omaha for his assistance and review of the sections about transition from college and working in academia.

My friend, John Gregorits, PE, a tireless champion of professional engineering and the NSPE; and an unforgettable mentor whose enthusiasm, sage advice, and encouragement will always reside in my memory.

All my supervisors, coworkers, and mentors over the years who have imparted their wisdom and advice and taught me by setting wonderful examples.

The NSPE community, at all levels, for providing me with abundant opportunities for many kinds of personal and career growth.

When I graduated from engineering school, my engineering universe was not remotely similar to yours today. The slide rule remained the stalwart companion of every engineer. Although it is obsolete and almost unknown in function, or even in appearance, to many readers of this volume, I still keep mine as a link to those simpler times.

For readers of this work, technology is a major part of your life, furnishing significantly powerful work tools, communication systems, and even recreational activities. Technology is allowing you to do things that I could not even imagine in my wildest dreams as an engineering student or apprentice engineer. The advent of the microprocessor was very exciting for the old-timers to watch. The applications today seem to be advancing at blinding speed. I recently heard that a major electronics chain is installing trash bins near the exits because what a customer just purchased is likely to be obsolete before it leaves the premises.

You are embarking on an exciting career. Adapting to new and constantly evolving technology will be challenging. Your commitment to life-long learning is of paramount importance to your success. Some of you will be working "shoulder to shoulder" with engineers around the globe because the Internet has made it possible. International collaboration is no longer a dream. It is as real as the continents and the oceans. We hope to give you some helpful pointers in this book, some of which may become as obsolete as the slide rule in a decade. But the personal development guidelines may prove to be timeless. If we can ignite even a small flame of desire to succeed, and if we can give you some help in fanning the flame, then we will have succeeded in our mission.

Just like my colleague, Doug Benner, I graduated from engineering school with no road map for career advancement. I must give proper credit, however, to my college professors for the fact that I left school with three specific goals:

1. I wanted to earn a master's degree in civil engineering while working at an engineering job.

2. I wanted to become a licensed professional engineer and, if possible, a licensed land surveyor.
3. I wanted to become engaged as an active member of one or more engineering associations.

Besides those specific goals, I also had a general goal to become the best engineer and land surveyor that I could. As the U.S. Army slogan urges, I wanted to be all I could be.

My career started in government service. My first engagement was as a junior engineering trainee in the Storm Drain Division of the City of Los Angeles, on the other side of the nation from my New England home and college. During that engagement, I took the Engineer in Training (EIT) examination and passed. As you know, the EIT (also known as FE) is the first step toward attaining licensure as a professional engineer. My next position was also in government as a commissioned officer in the U.S. Coast and Geodetic Survey, where I served my mandatory (yes, mandatory) military service.

Not until five years after graduating from engineering school did I enter private practice as a trainee in a highway engineering firm. Immediately, I began preparation for the Professional Engineering Practice Examination. I also joined the National Society of Professional Engineers and remained a member of the American Society of Civil Engineers, which I had joined in college. I passed the examination and started evening school toward a master's degree in civil engineering. I received the MSCE a couple of years later and also took the Professional Land Surveying examination, passing it.

So within the first decade following graduation, I attained my three specific goals. I didn't stop there, however, because I still had my long-term goal of success in the profession ahead of me. Now, a half-century from graduation, I find that my goal of achieving continued professional success is still there. My career in engineering has been delightful, personally and financially rewarding. Now, five decades after graduating from engineering school, my career is still going strong, continuing to give me great personal satisfaction.

One of the rewards of growing older is the opportunity to mentor—to help those who follow in our footsteps to succeed. That is our goal. Read this book with an open mind, pursuing areas of interest outside of its covers. Become your own champion. Engineering school is tough. You have some tools to use going forward. Use this book as a road map to your success.

Bernard R. Berson, PE, LS, F.NSPE

GETTING STARTED

1

TRANSITION FROM COLLEGE TO THE REAL WORLD

The party's over! It's time to get serious and start your career!

While you were studying engineering, you might have noticed that the business and liberal arts students had much more time to party than you had. You worked hard because studying engineering was difficult and demanding. So it's not really fair for anyone to tell you that "the party's over." But it is fair to say that when you start work, most likely you will work even harder than you did in school. Even if your work is not as intense as college, you will find the challenges, environment, and tasks to be quite different, which can sometimes make the transition difficult.

Please don't get the impression that you are going from the frying pan into the fire. The transition will be a great experience. The world of engineering work is characterized by a fast pace and very rapid change. In fact, we live in times of accelerating changes. That makes your new position exciting and challenging. Often, our news media teach us to focus on the negatives of our world. But our world is filled with good things; we live in wonderful and challenging times. Rapid technology advances have a very strong influence today; and you now have the fantastic opportunity to be a part of this rapidly changing age.

THE CONTRASTS BETWEEN COLLEGE AND THE WORKPLACE

It is important for you to understand the many differences between engineering studies and the engineering work environment. You will need to adjust your mindset accordingly.

Most changes in technology are due to the academic research environment. However, the route to the research environment usually requires acquisition of knowledge in a system steeped in the tradition of assimilating detailed and rigorous information in a cumulative manner. This process might be compared to assembling a kit with defined, sequential steps leading to completion. A key lesson to learn at this transition time is that the pace of work in the engineering work environment is often much faster than the one you experienced in college. This happens because of strong competition and the global economy. The college experience is carefully paced so that you can fully learn one block of instruction before moving to the next level. You have time to produce fairly orderly, perfect work. In the engineering work environment, there is often insufficient time to achieve the same level of orderliness and perfection. In practice, engineers normally are held to a legal standard of work that is not defined as perfection. You do the best you can in the allotted time, often trading elegance for safety, functionality, and cost.

Professors advance their career by being effective teachers, by performing rewarding research, by publishing books and articles, and by performing some consulting work. Engineers and their managers are motivated by designing competitive, revenue-producing products or projects to make their superiors look good and to earn challenging engineering assignments. In academia, a big motivator is the prodigious scholarly activity of research and publication. In private practice, the key motivator is satisfied clients (or taxpayers in the case of government or customers in the case of industry).

In college, you get some experience working on teams in your design, project, and capstone classes. If you go to graduate school, this teamwork continues, often with the beginning graduate student starting at the lowest level on the team. In academia, the key question about work is, "Does the work contribute to knowledge?" In the nonacademic environment, the key question is, "Does the work contribute to the business?" In academia, work is sometimes done primarily because it is interesting; in the nonacademic environment, it is done because it is financially worthwhile.

At college, you attend lectures, you read textbooks, you solve lots of problems, you participate in carefully structured laboratory exercises, you take tests, and you move on to the next level. It is a structured and protected environment. You are not as protected as you were in high school when you still lived at home, but you are more protected than when you go completely on your own to your first job.

When you transition to the workplace, you are finally on your own. It can be a scary time in your life. It is a good time to remember the old adage, "The more you learn, the more you realize how much you do not know." It is time to put your learning skills into high gear or, even better, put them into overdrive so that you can work intelligently.

LEARN THE CORPORATE CULTURE

My first reaction to that instruction was, "How do I recognize corporate culture when I see it?" Neither you nor I studied Corporate Culture 101. Here are some things to look for early in your career.

Corporate culture, like any other culture, is a set of behaviors and codes that guide how you interact with other employees and superiors within the company. Organizational culture can be described as the "personality" of a company. You can think of it as the composite of "how things are done around here." Corporate culture can be expressed in the interior design of offices, by what people wear to work, by the arrangements for hours of work, by how people communicate with each other, and by the formality of employee titles. It is attitudinal as well as structural.

A company's leader and senior managers shape the corporate culture profoundly. I recall that when the casual dress trend developed and many engineers discontinued wearing ties, an executive secretary told me that when the CEO in the corner office stopped wearing ties, I could do the same. Top management often defines the corporate culture. For example, if the CEO avoids conflict, other employees will tend to do the same, whether consciously or unconsciously. Behavior of senior team members—their reactions in a crisis and what they talk about regularly—all set the tone of the culture.

ENGINEERING CAREER AREAS

A wide variety of engineering employment sectors exist, and the transition from college to the engineering work environment may vary in different sectors. You will need to evaluate these differences as you journey through the transition experience. For some career choices, the transition will be complicated. This includes employment in industry, in government agencies, in research laboratories, in construction, in private practice, in consulting, in sales, or in management. Today, your entire working career is not likely to be in only one area. This means that there will be more transition experiences when you move from one engineering career area to another. Therefore, it is a good idea to keep a sort of transition diary of what works well and what does not work well; this will help you in future transitions.

STARTING OFF ON THE RIGHT FOOT

Here are some helpful suggestions:

- Be early to work; be prompt in the work environment. At college, if you violated these rules, you merely got behind the eight ball and had to play catch-up. On the job, violations of these rules can have more serious consequences.
- Focus on the quality of your work. Even though there may not be time to be absolutely perfect, avoid mistakes as much as you reasonably can.
- Hard work has multiple rewards: from your peers, your employer, and your clients. Other employees quickly note your energy level as well as your rate of learning. Don't wait to be spoon fed. Show lots of initiative. Show lots of interest. Make your manager a success.
- Expect difficult, complicated, and multidisciplinary problems.
- Make a concerted effort to understand how your work impacts the overall product or project. This will instill a strong feeling of the importance of your efforts. It also gives you insights into how you can expand the scope of your work.
- Learn to be business oriented. Learn the economics of your organization. Learn about your organization's competition.

- Define your short-term and long-term career goals, but don't assume every assignment must be strongly supportive of those goals. When an assignment is not strongly supportive, consider it to be a stepping stone on your career path. What are your career goals? How will you attain them?
- Continuously analyze and understand your personal SWOTs. What are your STRENGTHS? What are your WEAKNESSES? What are the best OPPORTUNITIES to enhance your career? What—and, possibly, who—are THREATS to your career advancement, and how will you deal with these?
- Develop your personal vision; this is very important. Your personal vision is discussed in the Leadership chapter.
- You are your own career manager; ask for regular feedback. The practice of your employer working with you to map out a career plan is over. You are the CEO of your career; you need to assume this responsibility starting on day one of your new job. Learn to work and network in the new multicultural and multinational environment. If you did not start to network with diverse people in college, this may be a daunting task. Even if there is limited diversity among your fellow employees, your exposure to teams and to clients will soon introduce the multicultural and multinational element.
- Be open to ideas from everywhere. Try to use your brain like a sponge and absorb as much as you can as fast as you can. Learn to network with as many people and resources as possible.
- Learn to be a team player, especially where the team must deal with conflict and team members seem to have different agendas.
- Have unyielding integrity. Ethics in the engineering environment is the subject of Chapter 11.
- Support your technical society and your university. Your technical society is a great resource for networking. Also, you need to give back to your university and to your profession.
- Seek a mentoring relationship. There will be more on mentoring in Chapter 20.
- Have fun! Find outlet activities that supplement and complement your career interests.

These are a few key suggestions to help in your transition from the student environment to the work environment. It is good to review them periodically throughout your career. You may even add to this list as you journey through your career. There is no magic formula for engineering career success, but paying attention to these concepts and suggestions will pay dividends in your future.

One of the by-products of your investment of time and energy in gaining an engineering degree is the refinement of your learning abilities. These will stand you in good stead throughout your career as you acquire new information in further schooling or at work. Protect these valuable learning abilities by using them frequently in your career and life.

2

LAUNCHING YOUR CAREER

In the previous chapter, we navigated the transition from your college experience to the working world of the young engineer. In this chapter, we continue that journey and focus on other aspects of your job and your employer that you need to investigate and understand as early in your career as possible.

If you set out to be a part of an athletic team that will become a championship team, you need to progress through several developmental steps. First you need to master the fundamental skills for one or more positions on the team. Then you need to develop a complete knowledge of all the rules of the game. Next you need to understand the overall strategy of the game as well as the strategy for each game situation. You need to be able to anticipate how your coach will act in each game situation. Prior to each game, you need to participate in analyses of the strengths and weaknesses of your opponent and the adjustments necessary to deal with these. After the game is over, you participate in reviewing the game tapes, learn from your weaknesses, and train to overcome these weaknesses. And the cycle continues.

Now apply this model to your job. Business is a competition. There are winners and losers. To survive, you need to work through similar steps. This chapter focuses on the rules and the strategy of the game. This book will help you navigate through all aspects of your job to help you become a part of a championship team.

EMPLOYEE HANDBOOK

If your employer is a large firm, there will be an employee handbook, which is an invaluable source of information. Don't yield to the temptation to put it on the shelf and use it only as a reference book. Read it from cover to cover. It is your rule book!

In a smaller firm, the employee handbook may be less formal and may even be just a series of policy statements. Many new and emerging organizations begin operations and grow without having a written employee handbook. The management of some new and emerging organizations consciously tries to operate the business without the "rules and regulations" of large organizations. However, as organizations grow, issues become more complex, often resulting in definition-of-policy memos or letters on select topics. For example, attendance control guidelines or benefits issues, such as holidays to be observed or eligibility for and scheduling of vacations, may be written to prevent misunderstandings.

The employee handbook is an essential tool for communicating workplace culture, benefits, and employment policy information to employees. What should be in the employee handbook? It typically describes information about the employer's employment practices, company benefits, equal opportunity commitments, attendance guidelines, pay practices, leave-of-absence procedures, safety issues, labor relations matters, and sanctions for misconduct. A carefully prepared handbook or manual serves an important purpose by helping to orient new employees, answer questions that arise during employment, and guide employment actions to comply with applicable laws.

It is likely that your employee handbook will define your firm's interpretation of *employment at will*. Employment at will is a traditional common-law perspective that says an employee may seek work and quit at any time and, likewise, that the employer may hire and fire at any time for any reason or no reason.

That sounds a bit wild and uncontrolled. That is why the courts in many jurisdictions have imposed certain limits to employment at will. For example, various laws include a provision stating that an employee is protected from discrimination or retaliation for exercising rights under the particular law. An employer's discharge of an employee for exercise of rights under law would be contrary to public policy, and in many areas, such discharge would be prohibited, thus imposing a limitation to the employment-at-will concept.

For these legal reasons, the employee handbook may define a set of reasons for which your employment could be terminated. Hopefully you will never have to deal with any such circumstances. If it happens that you do, you are advised to seek the counsel of a labor law attorney.

It should also be noted that the courts have held that an employee handbook or policy manual may be considered an implied contract if the manual contains certain promises relating to employment, tenure, or benefits. Consequently, the employee handbook becomes a de facto part of your employment contract. You may even be asked to sign a statement that you have read and understood it. If you are looking for a personalized employment contract, be aware that these are not customarily provided except for upper management positions with negotiated special terms and conditions.

ORGANIZATIONAL STRUCTURE

How is your organization structured? You are in an engineering group that is a part of a big organization, and the rest of the organization is just a whole lot of people that sometimes throw up roadblocks to your work, right? What is organization structure, really? It is the formal interrelating of individuals and groupings in allocation of tasks, responsibilities, and authority to achieve the goals of the organization. As you may have guessed, there are various kinds of company organization structures.

Corporate organization structures have been the subject of much discussion and research in the undergraduate business and MBA programs. Why, you may ask, do I need to know all this theory? It is certainly not mandatory for you to have an MBA to function as an engineer, but a knowledge of the basic concepts will (1) help you understand your organization better, (2) prepare you for the occasional restructuring within your own organization, and (3) help you understand how a competitor's structure can sometimes place it at a competitive advantage.

Government agencies and academic institutions also have hierarchical organizational structures but are less likely than corporations are to have many variations.

The traditional organization structure for many years was the pyramid shape with the corporate executive officer at the top with multiple management levels below and the production workers at the bottom of the pyramid. A miniversion of this pyramid can exist within the various functional groups, including the engineering department. It was generally assumed that to be promoted upward in this pyramid, you needed to gain years of experience and possibly even earn advanced degrees. Let's look at how the traditional pyramid evolved into a few key types of organization structures today.

Functional Organization Structures

Most organizations start with this structure. The organization has functional departments such as production, engineering, sales, R&D, accounting, human resources, and so on. It is a hierarchical, usually vertically integrated, organizational structure. It enables standardization within the organization and its processes, even for specialized employees such as engineers. There are advantages to this specialization. The organization develops experts in various areas. The product designers become very skilled in design, the production groups get flow diagrams down to a science, and the accounting department develops a great cost-tracking program. Individuals are very efficient in that they perform only tasks in which they are most proficient. This form is logical and easily understood by the employees.

The disadvantage of the functional organization structure is the difficulty in maintaining optimum coordination between functional groups. The groups tend to operate in their own specialized silos and often fail to coordinate with other departments. For example, the product may have an elegant design and may function for years before failure, but it turns out to be difficult to fabricate. The net result is a greatly reduced profit margin. This organization structure requires a strong and effective upper management to maintain interdepartmental coordination and produce organizational success.

Product Organization Structures

When the functional organization structure ceases to be effective, many companies convert to the product organization structure. Also, as new products and services are developed, the company may migrate to a product organization structure. Each product or group of products becomes a separate division or cost center. For instance, one division might make only television receivers, while another division might make digital video recorders/players. For the engineering firm, one division may specialize in residential site plan design and another in highway design. This divisional or product organization has the advantage that employees can focus on producing the specialized products or services for that division while still being able to use knowledge gained from other different, but related, divisions.

The disadvantage of the product organization structure is duplication of functions. Each division often has its own design, marketing, R&D, and other units that could be working together on common problems but do not under this structure. Extreme situations could occur where divisions actually compete against each other for the same customers. Also, each division often buys similar supplies in smaller quantities, therefore losing the opportunity for volume discounts.

Matrix/Project Organization Structures

Matrix organization structures are a hybrid of the functional and product organization structures. The organizations are temporarily modified to accommodate project teams and are therefore very suitable for engineering work. Employees from different functions form teams to work on a project until it is completed; then they revert to their own functions again. Each of the functions is meshed into each of the products (or product lines) to form a complex set of matrix relationships. Business decisions are usually made by the project leader and sometimes by the top corporate or top functional department head.

The major advantage of the matrix structure is that a wide range of skills are brought to focus on a specific project or product and the organization can move quickly for results. Because the manufacturing and marketing employees are involved with the design personnel in developing the product, it can be designed more quickly and produced and sold with fewer problems along the way. Problems can be identified before the design is finalized. Communications across groups that don't normally communicate is greatly enhanced for the betterment of all.

The matrix organization does have problems with reporting relationships, because employees may be reporting to more than one boss at a time. This can be confusing and create conflict for employees accustomed to the concept of having only one boss. When the function supervisor and the project manager impose conflicting demands on the employee, the employee's stress level increases and performance may suffer.

The matrix organization requires greater communication and negotiation skills and more adaptability on the part of employees. Employees must be assertive in dealing with the functional and product supervisors who may want to achieve conflicting goals. Employees must be willing to be flexible and to adapt quickly to changes.

Boundaryless Organization Structures

Some organizational development experts decided that even matrix organization structures are too slow and need to deal with change even faster in the new high-speed global competition. They developed the boundaryless—or network—organization structure. A boundaryless organization is an organization in which typical barriers and organizational boxes are eliminated and inside and outside organization units are connected in effective and flexible ways. Internal company barriers are removed between employees (departments are replaced by teams), and external barriers are eliminated by the company and suppliers working closely together as if all are parts of one organization. However, this is often more of a theoretical and conceptual goal than an actual organizational practice. Critics have called this the organizational utopia—it is never completely achieved in real life. Some large companies have tried this team approach without any formal barriers between employees and found that it is difficult to implement in practice. If it was ever implemented 100 percent, there would be serious control problems.

Modular Organizations/Networks/Clusters

Modular organizations have a core or central hub that coordinates the outside networks of suppliers or specialists that provide parts of the whole product or service. The network parts can be owned by the core company or by outsourced companies. The networks are added or subtracted as the need for their services ebbs and flows or changes. For instance, an athletic shoe company has a stable core function, but varying subcontractors in other parts of the world make its shoes. A personal computer company buys parts from various computer parts suppliers and assembles the parts from the outside suppliers at one central core location. Suppliers at one end and customers at the other end are brought in and made part of the organization; information and innovations are shared with all. Customization of products and services is made possible because of flexibility, creativity, teamwork, and responsiveness. Business decisions are made at corporate, divisional, project, and team levels.

The advantage of a modular structure is the minimization of the amount of specialization and specialists needed, as well as lower overhead. The company does not have to be all things to all people to produce a great product. It can outsource much and coordinate the product's assembly into a great product.

Disadvantages include concerns about what suppliers not under the core company's control might do. The suppliers may steal trade secrets, collude to raise prices, or terminate their relationship with little notice. Also, it may become difficult to tell where one organization ends and another begins. Coordination and communication skills become very critical in this organization structure.

YOUR ROLE IN THESE STRUCTURES

You will most likely find yourself in an organization formed like one of the structures just described. Actually, most companies are not strictly organized in one type of structure. Nor is one structure best for your career. It is important to recognize the differences, learn how to deal with them, and thereby optimize your chances for success.

In these organizational structures, you will usually find a series of engineering titles that lead to a department head or project manager position. Learn these titles and what is required to move from one to another. You will also learn that there are sometimes more than one series of career ladders, depending on the size of the organization. For example there can be a path or group organized by engineering discipline, by product line, or by project, or even as consultants/experts. Each may have a unique series of engineering titles. And, of course, there is the career path that leads to upper-management levels.

As you gain experience and receive promotions, you need to think about your career path options. You have to decide which career path meets your needs and fits your objectives. As you progress, you will need additional education and training. If you prefer technical work, seek more technical training. If you aspire to a management position, seek more business training. Consult with your mentor and/or your supervisor and do what is best for you.

Speaking of your supervisor, let's consider that position. Your supervisor implements directions from the top of the organization; enforces policy and procedures; has responsibility for the group's budget; and must deal with the problems, concerns, and technical issues of the engineers in your group. The supervisor has a difficult position.

Make your supervisor your friend, not your adversary. Learn what it takes to make your supervisor look good. Respect your supervisor's workload and respect your supervisor's need to be a "middleman."

In the product organization structure, your supervisor could be the program manager or the engineering department manager. The program manager determines what is to be worked on and the engineering department supervisor determines the best technical approach. The program manager wants the quickest, most economical job done, and the engineering department manager wants the best engineering job. Learn to live with these two sometimes conflicting sets of objectives and to negotiate compromises skillfully without constantly running to the supervisors to resolve conflicts. But as the old saying goes, never forget where your bread is buttered. Your reward system is discussed below.

MAINTAIN VISIBILITY

Recent graduates often assume that someone is looking out for them. If you set up a relationship promptly with a mentor, that may be true to some extent. However, you need to find out right away how your company is organized and who determines your raises. Don't leave your next promotion or raise up to chance. Regularly report your progress and results to the person who is responsible for your raises.

It is also important to do all that you can to gain visibility with upper management. If managers know who you are and what you have accomplished, you have a better chance for raises and promotions. If upper management knows who you are and that you have demonstrated good performance, they are less likely to include you if and when layoffs become necessary. You will also be picked for the better assignments and will be considered for special projects.

You can also enhance your career by developing good visibility with coworkers. When coworkers realize that you have good visibility with upper management, they will strive to have you involved in their projects and will seek your opinions on many aspects of their work.

The most common way to get good visibility is with presentations or reports to groups that include upper management. You must use these presentations to make management and coworkers aware of your skills and accomplishments for the benefit of the company as well as for your career. Your visibility has a technical element and a social element. Demonstration of your technical skills is the easy part. The social component includes your communication skills, your social interaction skills, and your team leadership capabilities. These are sometimes called the "soft skills" required for your career and will be discussed later.

There are multiple ways to demonstrate technical skills. Some are a routine part of your job; others are a result of extraordinary efforts. Here are some ideas:

- Writing reports
- Analyzing reports
- Attending courses
- Producing videotapes
- Developing simulations
- Fabricating models
- Filing for a patent
- Publishing articles
- Teaching a class
- Submitting a project or the team for an award
- Presenting a paper at a conference
- Writing a book

Here are some ways to develop the social component of your visibility:

- Attend social meetings
- Participate in intramural team sports
- Attend group luncheons
- Select prominent seat locations at meetings
- Volunteer for community programs
- Become active in a professional society

- Host a society meeting
- Socialize and network at conferences
- Participate at the company picnic and other functions

Remember, if you develop good visibility and use it well, you develop power and a better chance for career success.

YOUR REWARD SYSTEM

Two questions you should answer as early in your career as possible are:

1. How is my performance evaluated?
2. What aspects of my performance are considered most important by my employer?

Most larger organizations use both a formal scheduled performance appraisal and informal periodic feedback. Because formal performance appraisals require a lot of work and sometimes become confrontational, many supervisors postpone them as much as possible. Let's look at performance appraisal systems and what makes them work well.

A positive understanding between the employee and supervisor regarding job performance is essential for all employee and management work relationships to survive. "How am I doing?" is often one of the most urgent questions on your mind.

An effective performance appraisal system requires three key ingredients:

1. Your job description defines what is expected of you in the performance of your job.
2. Your supervisor provides frequent informal feedback about your work.
3. You receive an annual (or more frequent) formal performance appraisal.

The performance appraisal gives feedback in a constructive, coaching, and mentoring relationship. The performance appraisal is the supervisor's judgment of how well you perform your job based on established job measurement criteria and previously established goals. Ideally, your job performance is being appraised and rated, not your personality traits. Decisions

related to promotability, advancement, selection for training, salary administration, discipline, and even potential termination may flow from the results
of an objective performance appraisal process.

Ask your supervisor to provide a description of your organization's performance appraisal system. It should include the following key components:

- Criteria for job performance must be related to the job itself and
  reflect accurately and realistically the unique requirements of your
  position, level of assignment, and operating conditions.
- You should participate directly in establishing the achievement measures for your job. To do a better job, you should know what is
  expected, how you are doing on the job, and where assistance can be
  obtained when needed.
- Your supervisor is responsible to coach and collaborate in your development. Your supervisor should recognize and be concerned with
  your personal aspirations, motivation, and career growth needs.
- Your supervisor should communicate job standards and other expectations to you before the evaluation period begins. Then you will
  know what constitutes good performance, and your supervisor can
  more objectively assess performance.
- Effective compensation systems must link performance achievements
  to salary increase considerations. Even the best compensation plan is
  difficult to administer if performance is not the major determining
  factor in granting salary increases.
- Frequent feedback sessions should be conducted with you throughout the evaluation period. Regular "mini-session" reviews help minimize your being surprised at appraisal time.
- Career path formation and counseling must be a part of the performance review cycle. By focusing on performance accomplishments,
  you can receive more precise guidance about career options with the
  company and the achievement of agreed-upon milestones.

After you test your organization's performance appraisal program
against these criteria and you find it to be deficient in one or more areas,
what do you do about it? Actually, most performance appraisal programs do
not function as well as the theory says they should. Also, the fully functional
system just described often exists only in larger organizations. Nevertheless,
you should evaluate the deficiencies and try to work around them. For

example, if your supervisor gives you very vague standards and performance criteria, generate your own and seek your supervisor's concurrence.

Regarding the first and second items above, keep in mind that these are ideal components. You will learn a laundry list of measures of your performance. Now you need to learn which of the items on the list are most important. If you are fortunate, your supervisor will tell you. Otherwise, you need to evaluate the rankings constantly by observing the kind of routine feedback given to you as well as to your peers. And don't forget that those who evaluate your performance are human and will always tend to use informal criteria.

Informal evaluation criteria are hard to define and are subtle factors based on personal biases or beliefs of your supervisor. Is your dress, grooming, and appearance compatible with that of your supervisor? Is the tidiness of your office more or less compatible? You may be happy with a messy desk, but if your supervisor is a neatnik, that situation may be detrimental to your appraisal. Are you a people-oriented or results-oriented type; how does that compare with your supervisor? Do you carefully plan your work so that there is no last-minute putting out fires? How does this compare to work planning by your supervisor? We are not implying that you need to change your personality or that you need to acquire your supervisor's negative traits. We are saying that you need to recognize these differences and convince your supervisor that "your way" helps your performance.

There are two other possible human resource systems that are related to performance appraisals. They are the salary administration program and the organizational development program. Both tend to exist only in larger organizations. Both of these are beyond the scope of this book for in-depth analysis.

The salary administration program, as well as your department's merit increase budget, defines a range of possible salary increases available to you. Make sure your increase is commensurate with the performance level determined for you. The merit increase budget is a part of your department's labor budget.

The organizational development program, sometimes called a succession plan, is used to determine positions for which each employee has the potential for promotion, both in the short term and in the long term. Essential information is recorded from performance appraisals concerning the strengths and weaknesses of each employee in relation to career development, training needs, potential for advancement, and suitability for other positions.

USING THE
COMPANY TRAINING PROGRAMS

Your performance appraisal may have identified some training needs. You may have some training needs because of working in a new technology, because you need to learn a new computer program, or because you would like to transfer to another type of engineering work. You may need training to meet continuing education requirements for your professional engineering license. For general career development, perhaps you would like to earn an advanced degree. What are the options in your organization, and how do you learn about them?

Most companies support training programs and further education for their employees. Training exposes the employee to new and improved methods, improving productivity and reducing costs. Training exposes the employee to new contacts that can provide information about customers and the competition. Often the employee returning from training can return to the job and transfer knowledge to other employees.

Corporate and business training is a very large industry estimated by the American Society for Training and Development (ASTD) to be about $109 billion per year in the United States, with a $29.5 billion portion being spent on external services. Simba Information's strategic market report, *Corporate Training Market 2005: Forecast & Analysis,* projects global corporate training sales of U.S.-based firms will be $10.72 billion in 2005.

ASTD is the organization of professionals who do corporate training. There are about 70,000 members of this organization worldwide. The *ASTD 2005 State of the Industry Report,* from its Web site, indicates the following:

- The annual training expenditure increased to $955 per employee, up from an average of $820 per employee in 2003 and 2002.
- Employees are receiving more hours of formal learning—32 hours of learning per employee in 2004, up from 26 hours in 2003.
- Training delivery via learning technologies increased to 28 percent in 2004, up from 24 percent in 2003.

If the training you desire was not identified in your performance appraisal, you may have to sell the idea to your supervisor or to those at higher levels. This selling process can be an important part of your maintaining visibility. Keep in mind that your supervisor may resist because of insufficient funds in the budget or because your supervisor needs you on the job to meet deadlines. If so, be patient and work out a plan to have the training later when conditions are better.

Be aware that successful completion of training does not provide a guarantee of career advancement or promotion. You need to use the training to improve your performance. Also, it is possible that even with improved performance you will be unable to receive a promotion because a higher-level position is not available. If that happens, you should consider transferring to another department where a higher-level position is available. Or you may even have to change companies to receive the advancement you deserve.

Training or continuing education can be in one of three major categories:

Internal training programs. Larger organizations may have a large number of these programs. They may be conducted during or after working hours, which means adjusting your plans for personal time. They may be technical courses conducted by local college professors, by vendors, or by experts from within your organization. Alternately, they may be soft skills courses such as report writing, public speaking, negotiations, supervisory skills, and so on.

The most effective internal or in-house training courses are those that teach you to use existing company resources to get your job done more quickly, more easily, and with improved quality. These are courses like computer training, computer-aided design (CAD), computer-aided manufacturing (CAM), and engineering design programs.

External training programs. There seems to be an unlimited supply of external training programs. Today there are multiple delivery methods for these courses, including off-site seminars or symposiums, Web-based courses, satellite-delivered courses, correspondence courses (now available with multiple DVDs), and a plethora of tutorials.

For many years, external training programs required employees to be uprooted from their desk and daily work scenarios and placed into a classroom or some other kind of structured learning environment off the premises. Today, e-learning technology is making training much more seamless. Computer programs can prompt an employee to take a quick tutorial right at a workstation if the system perceives that a task could have been executed more efficiently.

Because corporate training is a competitive industry, engineers regularly receive training solicitations by e-mail and postal mail. Before you select a training course, you and your company experts should seek an unbiased evaluation of the course. The course content may fall short of the claims of the wonderful rhetoric in the brochure. It is recommended that you use a reliable and well-known training organization. The engineering technical societies have a large number of both technical and nontechnical courses with a history of good results.

Tuition aid programs. Large organizations often have tuition aid programs either for taking a few specific college-level courses or for the pursuit of an advanced degree. It is more difficult for your company to justify payment for an advanced degree, both because of the high cost and because of the higher risk of your seeking employment elsewhere after you complete it. Nevertheless, some organizations are so convinced of the value of the advanced degree that they even give time off with or without pay to study.

You need to evaluate your organization's tuition aid program carefully. There are many variations in these programs. Some will require you to pay a part of the cost. Some will use an incentive program that pays a portion of the cost that increases with your grade on each course. You should evaluate the necessary paperwork and approvals. You should also investigate the tax implications of tuition aid reimbursement. (For more information, see *www.groco.com/readingroom/tax_educationexpense.aspx*.)

Finally, you need to build a "business case" to help you gain approval at multiple levels for your tuition aid application. An important resource to help you through all these efforts is the advice of others in your organization who have gone through the process.

Should you study for a master's degree in engineering or an MBA? It depends. That old weasel-word answer works well here. Many who have paired their engineering education with an MBA are glad they did because it makes sense—good business sense. However, engineers who intend to follow a research career path or a highly technical career path will disagree. The correct answer depends on what you want to do with your career.

It is important that you analyze your current personal situation and your job situation when you are considering whether to enroll in a master's degree program. It is not smart to base this decision on the economy alone, as that is too unpredictable from year to year. Unfortunately, this makes it nearly impossible to accurately forecast the return on investment of your master's degree. If your employer pays for all or part of your master's degree program, this greatly changes the cost-benefit analysis. However, you still need to evaluate the intangible costs, such as the loss of significant personal time, while you are involved in the graduate program.

Master's degree in engineering. If you prefer to minimize your career involvement in management, this degree may be the best choice for you. If you pursue a technology master's degree, you will select a specialty subject and become an expert in this technology. In so doing, you may get one step ahead of the current trend—that a five-year degree is a requirement for the professional engineer's license.

The American Society of Civil Engineers (*www.asce.org*) is a pioneer of this concept, with its ASCE Policy Statement 465, adopted in October 2004. (*www.asce.org/pressroom/news/policy_details.cfm?hdlid=15*). This policy statement includes the following:

> Admission to the practice of civil engineering at the professional level means professional engineering licensing requiring attainment of a Body of Knowledge through appropriate engineering education and experience. Fulfillment of this Body of Knowledge will include a combination of:
>
> - a baccalaureate degree;
> - a master's degree, or approximately 30 coordinated graduate or upper-level undergraduate credits or the equivalent agency/organization/professional society courses providing equal quality and rigor; and
> - appropriate experience based upon broad technical and professional practice guidelines that provide sufficient flexibility for a wide range of roles in engineering practice.

This proposed change in engineering requirements is due to the fact that the civil engineering profession is undergoing significant, rapid, and revolutionary changes that have increased the body of knowledge required of the profession. Primary reasons for the body-of-knowledge increase are the rapid changes in technology, the effects of globalization, the increasing diversity of our society, increasing environmental concerns, and the concept of sustainable development.

The current model law of the National Council of Examiners for Engineering and Surveying (NCEES) recommends one year of experience credit for a master's in engineering. Check your state regulations to determine if this applies to you. Having recently completed the master's in engineering degree, you have polished your problem-solving skills and reviewed engineering theory, so it will be easier to prepare for the Principals and Practice part of the PE exam. The professional engineering licensure process is described in Chapter 17.

Another very recent proposed change to the NCEES model law would change the academic requirements for the professional engineering license to include a four-year degree plus 30 additional credits from approved courses. This proposed change is discussed in more detail in Chapter 18. Because it may influence your decision about a master's in engineering, you should monitor the progress of this effort.

The MBA degree. The MBA degree is generally believed to produce higher average lifetime earnings than does the master's degree in engineering. The MBA is recommended for those who expect their career path to move toward management. The MBA can also be helpful to those who select the technical career ladder.

However, as the number of MBA graduates multiplies each year, the degree loses a bit of the luster that it once enjoyed. This loss of reputation is also influenced by the increase in scholarly criticism of the traditional MBA programs that are so popular today. Traditional MBA programs are criticized as being too institutionalized, with an overemphasis on analysis versus leadership and interpersonal and problem-finding skills. MBA programs are also criticized for being too detached from the real business world, resulting in diminished improvement in usable career skills.

Nevertheless, MBA graduates are generally satisfied with their degree. The Graduate Management Admission Council® (*www.mba.com*) surveys MBA alumni who have been in the post-MBA job market for one to five years to find out what they think about their MBA degree and what it is doing for them in their career. In general, MBA alumni report higher salaries than before they earned the MBA, high rates of job satisfaction, confidence that they did the right thing by earning an MBA, and a sense of career opportunity that their MBA skills bring. It is interesting to note the kinds of industries that employed the respondents of this survey.

Products and services	38%
Consulting industry	18%
Technology industry	15%
Finance and accounting	13%
Manufacturing	7%
Energy and utilities	4%
Health care	2%

Note that it is logical for an engineer to work in all of these industries with the possible exception of finance and accounting.

As stated earlier, a training program or a graduate program is not a guarantee of promotions or higher salary. If you effectively apply what you have learned and improve your job performance, however, you will improve your chances for success in your career.

3

THE ENGINEERING PROFESSION

In this chapter, we define engineering and describe the functions of others on the design professional team. We also describe engineering licensure and the process to become a professional engineer (PE).

The concept of engineering, often oversimplified as "designing things," has been around for centuries. Some historians claim that the builders of the pyramids were the first engineers, or that Leonardo DaVinci (weapons, flying machines, etc.) was the world's first engineer. For practical purposes, you should be aware of the modern-day definition of engineering from the Accreditation Board for Engineering and Technology (ABET):

> Engineering is the profession in which knowledge of the mathematical and natural sciences, gained by study, experience, and practice, is applied with judgment to develop ways to use, economically, the materials and forces of nature for the benefit of mankind.

Engineering is a noble profession. It is about using materials, natural and manmade, and forces for the good of mankind. Added to this, ABET also includes formal study in its definition. Formal engineering education programs began to appear in the mid-1800s on several continents. In the United States, licensing of professional engineers began in the early 1900s. Today, the list of engineering disciplines and engineering specialties is ever growing due to constant advancements in technology. As observed by Gordon Moore, technology continues to grow and change at an accelerated rate:

> In 1965, Gordon Moore, cofounder of Intel, observed that the number of transistors per square inch on integrated circuits had doubled every year since the integrated circuit was invented in 1958. Moore predicted that this trend would continue for the foreseeable future. Today we don't count transistors but instead monitor data density, which has doubled approximately every 18 months, and this measure is the current definition of Moore's Law. Most experts, including Moore himself, expect Moore's Law to hold for at least another two decades.

All this increasing complexity leads to (1) more engineering work being accomplished by multidisciplinary project teams and (2) increased levels of design professionals working within the project team.

THE PROJECT TEAM APPROACH

Yes, some engineers work alone or in a small group solving problems, as in lab courses in college. But this is no longer very common. It is more likely that you will work with a project team. You may have had some experience with a multidisciplined project team in college.

The diversity of specialties within the project team can be categorized as follows:

- The more traditional engineering disciplines are civil, electrical, mechanical, and chemical. To these have been added industrial, aerospace, environmental, biomedical, computer, and many more. There are also many specialties within each of these disciplines.
- The engineering practice areas and economic sectors include construction, government, academia, industry, private practice, energy, agriculture, transportation, communications, consumer products, medical, and more.
- The functional specialties include research, design, development, testing, analysis, sales, and production.

The participants within the project team may include the following:

- *Designer or draftsman.* Creates technical drawings and documentation using CAD or other computer tools. These individuals have special training but are less likely to have college degrees.
- *Engineering technologist.* Supports the engineering function and lies in the occupation spectrum between the draftsman and the engineer. The engineering technologist usually possesses an associate degree or a bachelor of technology degree.
- *Engineer (whose role is defined above).* The engineer usually has a bachelor's or master's degree in engineering.
- *Research engineer or scientist.* Explores fundamental principles of chemistry, physics, and mathematics to overcome barriers to the desired outcome of the project. The research engineer or scientist often is educated to the doctorate level and often has one degree in engineering.
- *Architect.* Most often responsible for design of structures. Architects usually are graduates of a five-year or more academic program. How does the architect differ from the engineer? As you can imagine, that is not an easy question to answer. Wikipedia, the online encyclopedia, provides a succinct description:

> Planned architecture often manipulates space, volume, texture, light, shadow, or abstract elements in order to achieve pleasing aesthetics. This distinguishes it from applied science and engineering, which usually concentrate more on the functional and feasibility aspects of the design of constructions or structures.

The distinction between all these levels is often quite gray. In sports, you can "identify the players by using the scorecard." Things may not be that easy on a project team. It is incumbent upon you to know the following:

- Who performs in each role, as defined and accepted by the project team
- Who should perform what role, as defined by regulations and licensure
- Who is legally responsible for what aspects of the project

This is not an easy task. Ask questions of several team members and do some research on these issues. Regarding the second item above, the Web site of each state licensing board provides links to its respective regulations for the practice of engineering.

ENGINEERING LICENSURE

Starting with Wyoming in 1907, every state of the United States has enacted a statute governing and defining the practice of engineering to protect the public health and safety. You need to know what registration as a professional engineer means, how it operates, and why all engineers should register.

You will meet engineers who will offer multiple reasons to skip registration. Admittedly, many states provide exemptions under certain situations where engineers are not required to register.

Philosophically, modern society needs to regulate the practice of persons whose work involves the protection of life, health, safety, rights, and property. This clearly includes, but is not restricted to, the key professions of medicine, law, and engineering. Elimination and exclusion of the dishonest and unqualified from the practice of these professions are a matter of public welfare.

Registration effectively uses the law to maintain a clearly recognizable line of demarcation between the engineer and the nonengineer. With registration, the law enables the profession to maintain high standards of qualification and ethical practice.

Why Register?

Registration is the mark of a professional. The registration process demands an extra measure of competence, experience, and dedication. Registered engineers achieve an enhanced status in the eyes of the public, which can equate the engineers with professionals registered in other fields. Testimony in courts of law has increased credibility when it is provided by a registered engineer. Indeed, numerous documents, drawings, and applications require the seal of a PE to be acceptable at the local government level on up to the upper tiers of the federal government.

Registration demonstrates that you've accomplished a recognized standard. It sets you apart from others in your profession. Registration also provides career options and opportunities that might not have been available otherwise.

Registration is just the starting point for long-term professional growth and development. In the past, lifetime job security was a common employment condition. In today's business environment, however, with global competition, downsizing, and frequent job changes, the savy engineer realizes that the PE license can be indispensable for rapid transition to self-employment as a consultant or in his or her own engineering practice.

Finally, registration involves testing in engineering theory and engineering regulations. The longer you wait to take those tests, the more difficult they become.

How to Register

The process is fairly uniform among states. However, many states have subtle differences within their regulations regarding the licensure process. Fortunately, it is relatively easy to use search engines to locate the Web sites of the state licensing boards, which are comprehensive and helpful. Alternately, you are encouraged to visit the Web site of NCEES, the National Council of Examiners for Engineering and Surveying. You can think of this organization as the national association of all the state licensing boards. This Web site provides links with information about all state licensing boards and links to the Web sites of these boards.

4

ENGINEERING CAREER OPTIONS

Whether you are close to graduation or are already working as an engineer, throughout the course of your career you will likely work in more than one area of engineering. In this chapter, we explore the different assignments, sectors, and work areas of engineering to consider as either your first position out of school or for future opportunities.

Engineering career options can be classified by your specific engineering discipline as well as in the following ways:

Type of assignment
- Research
- Development
- Design
- Testing
- Manufacturing
- Sales
- Engineering management
- Teaching
- Construction operations
- Other

Major economic sector

- Automotive
- Computer
- Aviation
- Pharmaceutical
- Biomedical
- Environmental
- Agriculture
- Land development (residential, commercial, industrial)
- Public sector infrastructure
- Utilities
- Petroleum
- Plastics
- Food
- Military
- Transportation
- Telecommunications
- Aerospace programs
- Other

Type of work area

- Industry
- Construction
- Government
- Education
- Consulting or private practice

As you can see, there are many different combinations to consider, and you will likely hold a position in more than one area during your career because of your changing interests or new opportunities that arise. It is interesting to note that your career path could even lead you away from engineering. It is often said that your engineering training and logical approach to problem solving could lead you to a career as a stockbroker, attorney, actuary, independent entrepreneur, venture capitalist, or any number of alternate careers.

A plethora of information is available about engineering careers in various disciplines, engineering careers in various economic sectors, and about engineering work in the types of assignments listed above. This information is available on the Internet, from guidance offices, from engineering associations, and from the companies that comprise these sectors.

Within the organizational structure of the National Society of Professional Engineers (NSPE), the major types of work areas are called "practice divisions." The term *practice* is used in the same way as in other professions, such as law and medicine.

Engineering employment data is available from the U.S. Bureau of Labor Statistics (BLS) at *www.bls.gov.* The method of sorting data by the BLS does not match the career option lists used here. However, if you inquire by "Architecture and Engineering Occupations" and by "National Industry-Specific Occupational Employment and Wage Statistics," you can discover some interesting data. This can be useful to estimate where engineers of your discipline are employed and what are some average wages. The data is available on a national, state, and metropolitan-region basis. Keep in mind that the accuracy of the data is misleading because self-employed engineers are not counted. It was recently estimated that about 55,000 engineers are self-employed in the United States.

ENGINEERS IN INDUSTRY

Industry is the largest employer of engineers in the United States. We use the term *industry* to include manufacturing, telecommunications, and utilities. In industry, the engineer is likely to be assigned to an engineering department along with a team of other engineers from various disciplines. Engineers in industry may work in any of the types of assignments listed above. It is less likely that engineers will work in research in small companies, because small firms typically cannot economically justify a research function.

Unlike other professions, engineers in industry have many different elements to consider when producing a design or product. First, they are responsible for producing a good design or product at the lowest possible cost to meet internal budgets for their company and to produce a product worth selling. Likewise, they are also responsible for providing a safe, quality product or design at a competitive cost to the customer. Balancing these requirements, especially with a financially aggressive management, is discussed in Chapter 11. To be sure, this multiple-master dilemma exists in all engineering career options, but it has been proven to be stronger in industry.

Although many products are designed and developed by engineers employed directly by the manufacturer of the product, a sector of engineering employed by private firms offers product development services to industry. Such firms may be utilized when the manufacturer has limited design staff, when the specific skills required of the designers are not available in-house, or when developing the product or portion of the product with in-house resources is simply not cost justifiable. If product development outsourcing occurs, you may be asked to interface with the independent design firm as the engineering representative of your organization.

ENGINEERS IN PRIVATE PRACTICE

Engineers who work for consulting engineering firms are said to be in private practice. In this area, you may work for a larger, well-established firm or for a small, recently started firm. In either case, you are more likely to be entrepreneurial or to work with engineers who are entrepreneurial by choice. This requires a willingness to work long hours, an ability never to lose your enthusiasm, a willingness to spend part of your time selling your services, and a willingness to accept business risk. Are you prepared to be an entrepreneur? If you are not sure, you should review some of the abundant literature on this subject.

To work in private practice, you or someone in the firm needs to be a licensed professional engineer. The requirements and process for licensure are discussed in Chapter 17.

Many private practice firms offer combinations of civil, mechanical, electrical, structural, geotechnical, environmental engineering, and related services. But any engineer who sells an engineering service as a consultant, regardless of the engineering discipline, is effectively in private practice. Some consulting engineering firms offer a wide range of types of engineering design services. As noted in the prior section, some engineers in private practice offer to provide support services to industrial clients by providing product development support. That niche of engineering practice may have special contract and liability issues. (See Chapter 29.)

ENGINEERS IN CONSTRUCTION

Construction firms range in size from small residential builders to mega-firms that build bridges, skyscrapers, and major highways and operate on a multinational basis. Consequently, it is difficult to generalize about the type of engineering work involved. With a small construction firm, you may become involved in both design and project management for an entire project. For a very large construction company, the design function is often separate from the project management function, and each is performed by teams of individuals with various specialties.

The design-build approach to construction is a unique arrangement. The Design-Build Institute of America defines the design-builder as "the entity contractually responsible for delivering the project design and construction." The design-builder can assume several organizational structures. Most common would be a firm possessing both design and construction resources in-house, a joint venture between designer and contractor, a contractor-led team with the designer in a subcontractor role, or a designer-led team with the constructor in a subcontractor role. In the design-build firm, an engineer could be the designer or the contractor or may serve on a project team.

If you are an engineer who loves to be involved in construction and be involved from the conceptual design to seeing the project completely built, you should consider a career in construction.

ENGINEERS IN GOVERNMENT

Federal, state, and local governments employ many engineers. About half of these work for the federal government in agencies like the Departments of Defense, Interior, Agriculture, and Transportation, as well as in agencies such as the National Aeronautics and Space Administration (NASA). Most engineers in state and local government agencies work in highway and public works departments. Most towns and cities have a "city engineer" who may be on the city payroll or may be an engineer in private practice who is hired on a consulting basis.

Engineers in government are much less likely to do original design than those in other engineering disciplines. They are more likely to review and approve plans and designs by others. They will also act as advisors to the governing body and to boards and agencies within their jurisdiction. They may be asked to assist in the drafting of rules, regulations, ordinances, and standards for use within their jurisdiction. They may be called upon to provide input into capital improvement budgeting and to participate in short- and long-term capital improvement planning. They may be asked to participate in the acquisition of equipment, supplies, and services through the preparation of bid specifications and to evaluate the bids for compliance when they are received.

ENGINEERS IN EDUCATION

Engineers employed in education are primarily graduate students and professors. While a large university physical plant frequently requires an engineering function to build and maintain the facilities, these engineers are not educators.

Academic credential requirements usually specify that the engineer in education have a master's and/or doctorate degree. Especially in larger universities, the engineer in education both teaches and conducts research. The research usually results in publications and the development of very specialized knowledge. The specialized knowledge and advanced degrees allow consulting work for supplemental income. Professors often assert that their average salary is below that of their peers in nonacademic engineering employment, but consulting can fill that gap.

Engineers in education work in a culture that is very different from the nonacademic environment. Obviously, engineers in education need to enjoy teaching and research. Generally, their work is more intellectually challenging than that of nonacademic engineers.

DUAL CAREER LADDERS

Especially in industry, but also in other engineering disciplines, the concept of the dual career ladder has developed. It is based upon the assumption that, at some point during their careers, many engineers are asked to choose between engineering and management opportunities. In years past, the decision process was strongly influenced by the fact that engineers who accepted promotions into management had higher earning power than did those who maintained a technical path. Thus, some engineers who were really better suited for a purely technical career were force-fit into management. Today, astute organizations have alleviated this problem by providing a "dual-career ladder" concept. This concept established pay parity between senior level engineers and engineers who choose management positions.

Organizations that have established the dual-career ladder concept offer some nonsupervisory positions that require engineers to perform at a higher level, assume more complex duties, and use advanced technical skills. These positions can justify higher, management-level salaries for engineers while keeping the company's technical capability intact and consistent.

Of course, smaller organizations do not have the luxury of multiple positions to enable implementation of the dual ladder concept. If you are employed in a small organization and are asked to move into management against your better judgment, you can negotiate a better salary arrangement by explaining the positive results of this concept in larger organizations.

Bernie has had most of his experience in private practice, much of it in a small-firm environment. His experience was generally technical with concurrent management responsibilities. This combination was a great fit for him because he enjoyed the challenges of designing and problem solving but also felt a need to have leadership challenges in his career. Doug has had most of his experience as an engineer in industry with some assignments in information technology and human resources. Subsequent to retirement from industry, Doug had further experience in private practice and in construction. This diversity also fulfilled a desire for management and leadership experience.

You should seek a niche that is professionally rewarding as well as financially satisfactory. Individual needs and choices will dictate what employment roles you will pursue. If you are fortunate enough to find a mentor with sufficient years in practice to share experience and sentiments with you, this relationship will provide a significant source of assistance in forming your own vision for your future. Another valuable resource is the "career forum" section of the Web pages for various national engineering societies such as, for example, the one under the heading of "Forums" at *www.nspe.com.*

5

YOUR RÉSUMÉ

WHY YOUR RÉSUMÉ MATTERS

An impressive résumé is vital for launching your engineering career as well as carrying you through career changes. Your résumé performs important functions for you in a job search. It gets you through the screening process and to the interview. It helps interviewers know what to discuss with you during interviews, and it lets the interviewer know what you have done, what you think you can do, and what you think is important. It represents you when you can't be there, both before and after any interviews.

The time you spend in writing a good résumé is time well spent. Résumé preparation is an important skill to master; it is likely that you will use a résumé numerous times in your working career. A résumé is a self-marketing tool: a place for you to catalog your accomplishments and specify your experiences that are relevant to potential employer's needs. Even if you stay with the same employer throughout your career, you will need a résumé to represent your accomplishments to earn favorable performance reviews and raises and to compete for promotional opportunities.

The word *résumé* comes from the French word *résumér*, which means to summarize. So the exact purpose of a résumé is to summarize your experience, your knowledge, and your accomplishments. You must do this succinctly and in a manner customized to the potential employer's needs. Say exactly what you mean in the least number of words possible. Long résumés usually are not good résumés. Résumés should be from one to three pages long. A résumé that is too long will simply bore the reader. Among so much material, nothing will stand out and be remembered. A good rule of thumb: A recent college graduate's résumé should be one page, while a professional with 15 years of experience may have a résumé of two to three pages.

GUIDELINES FOR THE EFFECTIVE RÉSUMÉ

An effective résumé addresses the employer's needs and requirements and demonstrates a match between what you have to offer and those requirements. It stimulates interest in you by showing your unique qualities, your well-developed skills, your relevant work or academic experiences, and your accomplishments that clearly differentiate you from the competition.

Here are some key considerations for good résumé writing:

- *Tell the truth.* It is simply unethical to invent any part of your background, despite the fact that it seems to be happening more frequently. Remember, it is very easy for employers to verify the basic facts on any résumé, especially your title, dates of employment, prior salary history, and your major and degree. Most employers will dismiss employees, even years later, if they are discovered to have lied on their applications. Remember, your résumé, and everything on it, becomes part of your permanent personnel file.
- *Sell yourself.* Suppress your modesty. Employers will expect you to delineate the skills, abilities, talents, traits, and experiences you have relative to their needs. They will not expect you to hold back. You are doing them a favor if you put your best foot forward.

  It is often said that engineers are not good salespeople. If you are reluctant to brag, imagine what someone else, who likes you a lot, would say about you. What would your best friend, sister, mom, dad, or favorite professor say about you? That's what belongs on your résumé.
- *Be specific.* Stress accomplishments and achievements and give specific examples. It is difficult to provide many examples in a short résumé so emphasize a select few. But wherever possible, give quantitative results such as "completed the design three weeks ahead of schedule with a product cost 15 percent lower than originally estimated."

- *Show your technical depth and breadth.* In the engineering work environment, it is seldom deemed that a candidate was "too technical" for a position. Therefore, describe your technical strengths by emphasizing years of experience with the particular technologies included in the job description. However, don't provide a laundry list of all of the technologies that you have touched on in your career; employers may view this as ambiguous and unfocused.

- *Illustrate passion and commitment.* When choosing a candidate, an employer will often select someone who appears to be committed to the profession and passionate about his or her work. Highlight your involvement in your area of the engineering profession and your commitment to bettering yourself by listing licenses and certifications; by citing involvement in professional organizations; by listing articles published in professional journals and magazines; and by noting if you have self-funded any of these efforts. Commitment of personal resources is a clear sign of commitment to the field.

- *Make your résumé clean and crisp in appearance.* Use white space so the résumé does not look too busy. Single space within sections. Use boldface type to guide the reader's eye to important information.

- *Proofread carefully.* Be absolutely sure that the résumé is error free (no typos, no misspelled words) and easy to read.

These guidelines will help your résumé garner added consideration when being viewed by prospective employers. This small investment of your time will help you to "attain the key that unlocks the door" to your future in your engineering career.

WHAT TO INCLUDE IN YOUR RÉSUMÉ

Heading. Your contact information is important for obvious reasons. The first thing on your résumé should be your full name, at the top of the page, in bold type and in a slightly larger font than the rest of the résumé. Follow your name with your complete mailing address and zip code, telephone/cell phone number with area code, and your e-mail address.

Job objective. In one short phrase, tell the employer what kind of work you are looking for (e.g., "An entry-level position in mechanical design" or "A position in residential site plan development"). *Do not* use phrases that tell the employer what you want them to do for you (e.g., "A position that will help me develop my skills in construction engineering").

Education. If you attended more than one college, list them in reverse chronological order. Under each college, list courses that would interest the employer and support your job objective. Include any relevant academic honors. If they are not self-explanatory, describe briefly (e.g., "Tau Beta Pi, engineering honor society"). Recent college graduates should list education first. Engineers with ten or more years of experience should list their experience before their education; for them, experience is more relevant.

Experience. Include work experiences and significant projects in reverse chronological order. For each entry, on the first line put the name and location of the company or organization followed by the dates of employment or participation. List the job title on the second line. Then describe your general duties and responsibilities. Next describe an accomplishment that you are proud of. Use short phrases beginning with action verbs to describe your experience. Use present tense verbs for current jobs, past tense verbs for previous jobs. Don't use personal pronouns. Leave out phrases such as "Responsible for . . ." and "Duties included . . ." and the headings "position," "job title," and "duties."

Skills. List technical skills and other skills that will support your job objective and fit the employer's needs. For example, you might include design or analytical skills, foreign languages, laboratory skills, teamwork, or leadership skills. Always include your relevant computer skills. Cite accomplishments and achievements as explained above.

References. Include on your résumé only if needed to fill the page. Otherwise, prepare a separate page for your references. Provide three or four references—at least one, preferably two, from previous employers; the remainder from faculty, your advisor or department head, or others who are familiar with your work habits. Do not use personal references. Include name, title, address, and phone number for each reference. Be sure your name appears at the top of the reference page in case it gets separated from your résumé. If you have room, include "References available upon request" at the bottom of your résumé. This book includes two sample résumés for engineers (see the Appendix).

THE SCANNABLE
AND ELECTRONIC RÉSUMÉ

Some employers are now requesting scannable résumés. Instead of being read by a human, the résumé is scanned with an OCR (optical character recognition) scanner and stored as a text image. Also, more employers today are requesting and receiving electronic résumés from Internet résumé services or by e-mail directly from applicants. Because the process of scanning hard copy résumés is time consuming and not as accurate, electronic résumés are preferred.

With either scannable résumés or electronic résumés, a keyword search is performed to identify candidates with the skills defined in the job description. Keywords are nouns or noun-based phrases that describe your skills and accomplishments. Keywords for a computer engineer might include *C++, Ada, AutoCadd, Windows, HTML, online applications, and information systems management.* The résumés with the most matches are likely to be selected for the interview.

If you want your résumé to be selected, it is essential that you understand the requirements of the job for which you wish to be considered and that the keywords that apply to that position appear frequently on your résumé. They can appear in your objective, course descriptions, skills listings, and descriptions of your projects and work experiences. It appears that résumé selection by keyword hit count is a poor substitute for the human evaluation of the quality of the work experience. However, keep in mind that human evaluation, if performed by someone without experience in the particular job that is open, can be just as inaccurate. The résumé selection process is one of compromises.

To optimize your chances for success, you might also include a "keyword summary" section on your résumé. It should follow your objective and include keywords (and their synonyms) that apply to the type of work that you are seeking. A keyword summary should not be necessary for larger hiring organizations, which have sophisticated keyword search algorithms in their résumé-processing equipment.

The format for your scannable résumé may be slightly different from the résumé you have prepared for the human reader. Many of the elements that you used to draw the human reader's attention will have to be eliminated from the scannable résumé. OCR scanners and their associated reader algorithms aren't very bright. Often they can't read the significance of bullets, italics, and underlined words used to emphasize your accomplishments. Similarly, bold, shading, and fancy fonts are likely to create confusion. Use the following tips when formatting your scannable résumé:

- Use capital letters to emphasize important points.
- Avoid using horizontal or vertical lines on your résumé.
- Use lots of white space.
- Use standard, easy-to-read fonts like Times New Roman, Courier, Arial, or Universal.
- Use 10- to 12-point font size; don't use compressed fonts or compressed spacing.
- Don't worry if your résumé is two or even three pages long if it is to be scanned; the computer doesn't care as long as the text is easily scanned.
- Use plain white or light-colored paper; mottled or granite papers will confuse the scanner.
- Most scanner programs will use "artificial intelligence" to extract important information and store it in a database. To avoid confusion, the first text on your résumé should be your name, followed by your address, phone number, and e-mail address, with each element appearing on a separate line.
- When you mail your résumé, send it flat and unstapled in a large envelope.
- If you are faxing a résumé that will be scanned, be sure to set the fax machine on "fine" mode instead of "standard" mode. Faxing your résumé will take longer, but you will increase the likelihood that it will be scannable when it reaches its destination. Play it safe, though, and send a hard copy as well.

THE CURRICULUM VITAE

The curriculum vitae is often required when promoting yourself within professional and academic fields, for those applying for professional programs, for those applying for employment with international firms, for those applying to graduate programs, for inclusion in engineering proposals, and for inclusion in engineering brochures. It is also used to provide introductory material for individuals who make speeches. You should start compiling data and maintaining your curriculum vitae early in your career, even though you may have no immediate need for it.

The curriculum vitae differs from a regular résumé in that it is as long as needed to define your education, experience, and other information about your accomplishments. It can be customized to delete irrelevant information for each use, but it should include:

- Professional, vocational, or research objective
- Summary of qualifications
- Professional licenses or certifications
- All college education
- Listing of relevant coursework to match career or academic objective
- Educational or professional honors or awards
- Project and work experience, including quantifiable results
- Scientific or academic research, laboratory experience, and related skills
- Description of thesis or dissertation, papers written, publications
- Technical and specialized skills
- Professional presentations
- Professional and association memberships
- Community, civic, and religious involvement
- Travel/exposure to cultural experiences
- Foreign language skills
- Additional information that may support objective or qualifications

RECORD KEEPING

Be sure to keep careful records of your job search, including all your applications, the associated customized résumé, all correspondence, and the results. You certainly don't want to send a customized résumé, then go for a job interview three weeks later and forget exactly how you customized the résumé. A curriculum vitae can be used as a method to help organize your data for record keeping purposes. In addition, it is a good idea to keep notes about the good, the bad, and the ugly of the interview.

Also, between job searches it is important for you to keep careful records of your positions, your accomplishments, your performance appraisals, your continuing education, your licenses and certificates, your publications, your patents, your copyrights, your job changes, your community activities, and anything else that can be used in preparation of future résumés. Wow, that can be a lot of material! And it should be kept in a safe place, not in a shoe-box.

I recommend that you maintain this information as hard copy files and electronically. As an added resource, a free template service called CAM Index (Career Asset Manager) is available at *www.ieeeusa.org/careers/cam/* that can help you organize your records.

6

INTERVIEWING SKILLS

When you leave college, you will need to interview for your first job. Many engineering students participate in co-op programs or internship programs so they start to interview as early as their sophomore year in college. There will be many other interviews throughout the course of your career for various reasons, whether you are changing positions within the same company or moving to a different company. No matter what the purpose, the interview is a forum to market yourself and your accomplishments. Most engineers are not very interested in sales and marketing and are not very good salespeople. Nevertheless, because the interview is effectively a "sales call," you need to learn how to present yourself and your accomplishments well to ensure career success.

During the course of your career, you may experience job interviews, promotion interviews, performance appraisal interviews, organizational restructuring interviews, and the like. This chapter focuses on the job interview, but the information here can be readily modified for other types of interviews.

Interviews may be structured in a variety of ways:

- *Formal interview.* The formal interview is held in an office or conference room setting by a representative of the human resources department or the hiring manager. The formal interview uses a well-structured set of questions.

- *Informal interview.* The informal interview may take place anywhere. It may be an interview by telephone. It may or may not be well structured. It is often not planned and is used under special circumstances such as at cocktail parties, at lunch, or at a professional society meeting. An informal interview may pose a risk for the employer because the lack of structure may cause violations of government standards.
- *Planned interview.* In the planned interview, the interviewer has a plan of action with prepared comments, prepared questions, prepared alternate questions, and a time schedule in mind. Obviously the plan must include some flexibility, but the most effective interviews are carefully planned.
- *Nondirective interview.* The nondirective interview is freewheeling and designed to let interviewees speak freely. The interviewer must be skilled in keeping the interview on track without interviewees being aware of it. The interviewer must be a careful and patient listener. This interview is designed to give candidates complete freedom to "sell" their abilities.
- *Depth interview.* The depth interview is intended to thoroughly examine the candidate's background and thinking and to evaluate the engineer's ability in his or her area of specialization. This type of interview is used to evaluate the candidate's claims of technical expertise.
- *Stress interview.* The stress interview, which is relatively uncommon, is designed to test the candidate's character and behavior by imposing conditions of stress and strain. These types of interviews are controversial but are sometimes used to test the behavior of individuals who will work under disagreeable, trying situations. A variation is the "in-box interview;" a simulation exercise where the interviewee is given an in-box containing a series of routine and stressful problems.
- *Behavior-based interview.* Behavior-based interviewing is a thorough, planned, systematic way to gather and evaluate information about what a candidate has done in the past to show how the person would handle future situations. It is accomplished by carefully developing questions that focus on the key competencies for the job. Each candidate is then asked the same questions. Interviewers look for consistency of responses and things that are comparable between candidates—things that can be scored. For example, if the job requires you to perform multiple tasks and switch between them, the

interviewer will want to learn about any experience you had like that in the past.

- *Group interview.* The group interview is arranged so that several interviewees are assembled and observed in a group discussion arrangement. The objective is to observe how the candidates react to and against each other. This kind of interview is rarely used.
- *Panel interview.* The panel interview is conducted by a formal or informal selection committee and is sometimes done for managerial positions. Panel interviews have the advantage of using the collective judgment and wisdom of members of the panel and, theoretically, select better candidates.

The formal interview and the planned interview are most commonly used. The depth interview is used to evaluate technical expertise and accomplishments. The other interview types are rarely used.

WHAT TO EXPECT IN THE INTERVIEW

The typical interview is composed of several steps. The first step is a brief introduction and establishment of rapport. During this time, the interviewer observes your overall appearance (dress, body language, speech, professionalism) and learns about you as a person. There is a discussion of your career objectives and how this melds with what the employer is seeking. Then there is a review of your credentials and your experience. Next, there may be a discussion about the job opening and the organization. Finally, you'll have an opportunity to ask questions of the interviewer. As the interview wraps up, you will be told what will happen next.

PREPARE FOR THE INTERVIEW

If you properly prepare for the interview, you will appear to have confidence and control of your portions of the interview. Go into an interview with the feeling that you will impress them so much that they will have to make you an offer. Early in your career, it may be difficult to muster up lots of confidence because your résumé seems barren. Don't let that bother you. Use your college activities, extracurricular activities, summer jobs, internships, co-op experience, and even high school achievements to bolster your confidence. Here are some other hints for proper preparation:

- *Research the organization and the position.* This can be viewed as a due diligence exercise. *Due diligence* is an expression that has legal significance but more generally means verifying data, reviewing documents, and "running the numbers." Find out the organization's products and services, annual sales, structure, and other key information from Internet search engines, the organization's Web site, the organization's brochures, the public library, a stock broker, professional magazines, or from former employees. Show that you are interested in working for the prospective employer by demonstrating knowledge about the organization.
- *Use your research to develop a set of questions.* Limit and focus your questions to what will help you determine whether this is the job and the organization for you. Don't overpower the interviewer with frivolous questions about details that are not germane to matching you to the position.
- *Bring extra copies of your work samples (e.g., engineering calculations) or project reports.* It's easy to say you can do a particular kind of engineering work. It's better to give examples of work you have done. Anticipate technical questions you may be asked based on the requirement of the position you are seeking. Think of recent strong examples of related technical work you've done; then when the question is asked, answer with specifics including work examples.
- *Be prepared to discuss your strengths and weaknesses.* Think about these in advance, especially as they may apply to the open position. Only you can recognize your most valuable strengths and most vulnerable weaknesses. Prepare to emphasize your major strengths. Your weaknesses, if they must come up, should be tactfully turned around to positives.
- *Conduct practice interviews.* Get a friend, a list of interview questions, and a tape player or video recorder and do an interview rehearsal. If you are still in college, work with your career services office. In your rehearsal, include a presentation or demonstration if that will be part of the interview. Start with introducing yourself and go through all the expected interview steps. With your friend, analyze the recordings and your observed performance. Write down all

areas that need improvement. Write down all questions with which you had difficulty. Practice until your rough spots are worked out and your delivery is smooth.

- *Dress appropriately.* Proper grooming is a no-brainer, but proper dress may require some thought. The interviewer's first impression is very important. Dress in conservative clothing, select subdued colors, use limited jewelry that does not look cheap, wear conservative accessories, and be sure everything is clean, neat, and professionally pressed.
- *Bring extra copies of your résumé on résumé-quality paper.* And don't forget names of references and letters of recommendation if you have them available. Bring all your material in a professional folder or binder.

AT THE INTERVIEW

To increase your chances of success, follow these steps:

1. *Arrive 15 minutes early.* Bring all your information to complete a possible employment application.
2. *Show enthusiasm.* Interviewers are impressed when a candidate exudes enthusiasm for the job. The assumption is that the enthusiastic candidate will be highly motivated on the job and will be an eager learner.
3. *Act professionally.* This means you need to communicate intelligently, use proper grammar, make eye contact, listen, and ask intelligent and relevant questions. If you concentrate on acting professionally during the selection process, you will have an advantage over other job seekers. The hiring manager will tend to choose the most professional applicants among otherwise equal candidates because the manager believes that person will represent the organization well.

    If time permits, you should review the basic theories of body language. Some interviewers use body language and facial expressions to evaluate you and your responses. For example, maintain eye contact, sit straight, do not put elbows on the desk, and do not cross your arms. Be sure to use a strong handshake; three pumps is the norm.
4. *Be honest and show integrity.* Be candid and open about past jobs. If there are any negative aspects, present them in a positive light. For example, if you were terminated from a prior job, be truthful without

being negative and highlight your strengths or how you learned from that situation.

5. *Avoid being evasive when asked difficult questions.* For example, if you don't have a requisite skill, admit that fact. Don't try to cover it up with lots of words, and don't give examples that aren't relevant. You should point out that you do have some related skills and that you're always eager to learn.

6. *Avoid being too wordy.* Responses should be no longer than a few minutes.

7. *Ask pertinent questions.* You developed a set of questions during your preparations for the interview. Use some of them at appropriate times during the interview. Also, save some for the end of the interview, when you are frequently asked if you have any questions. You may even develop a tough question about the organization; this shows that you have given employment in the organization some serious thought.

8. *Say "Thank you."* Thank the hiring manager or human resources representative for the interview and for the person's time. Ask for a business card so you can follow up with the interviewer.

9. *Follow up.* Send a thank-you note via e-mail or regular mail.

HOW ARE YOU EVALUATED?

In addition to matching your education and experience to the needs of the organization, the interviewer evaluates your nontechnical skills. Here are some of these key skills and how you can prepare to demonstrate competency in each of them.

- *Organizational skills.* These are essential for any engineering position. Employers can get a sense of how an individual will handle large workloads by how organized that person is during the interview. The engineer who makes a strong effort to stay organized during the interview is an employee who will take a job seriously and make a sincere effort to get things done.

- *Critical thinking skills.* Display your critical thinking skills and show your ability to think on your feet with your responses. Create the impression that you are a problem solver. How can you practice these skills? Prior to the interview, prepare of a list of situations from previous jobs that required critical thinking to solve a problem. Be prepared to use these examples in the interview. During the interview, talk your way through the answers. Make sure the interviewer can follow your train of thought when responding to questions.

- *Communication skills.* Engineers often lack self-confidence in communication skills, and engineers often fear public speaking. "But I studied engineering, not journalism or political science," you may be saying. True, but unless you can effectively communicate ideas during the interview, you may not come across as very confident or very competent. This is why engineers are expected to have good communication skills, often including public speaking, even though early in their career they may have limited exposure outside the engineering department.

  As previously suggested, practice speaking—or answering interview questions—in front of a recording device. This will get you accustomed to speaking aloud and let you observe the areas where you can improve your presentation skills.

- *Interpersonal skills.* Engineers need to work well in a team environment. You need to display interpersonal skills in the interview. You need to be receptive to the ideas of others. You need to convince the interviewer that you will work hard on your own and that you will be a positive contributor to the effectiveness of a group effort.

  Here again, preparing a list of examples is a good idea. If possible, use an example where you were the team leader. Also, be prepared to discuss troubles that you had to overcome within a team.

- *Multitasking skills.* Your interview should demonstrate your multitasking skills. As your engineering career progresses, you will find yourself participating in multiple projects simultaneously. Your ability to prioritize your time and effort among several projects is important. The way you respond to multiple-part interview questions can reveal your multitasking skills. Also, when discussing previous positions held, include situations where you worked on multiple tasks simultaneously.

- *Dealing with stress.* Be prepared to deal with questions that are uncomfortable; such questions may be a subtle technique to measure your skill at dealing with stress. This is not a widely used technique, but it can happen. Here are two examples:

  > *How did the realities differ from your expectations in your previous job?* This was probably not on your list of possible ques-

tions; that is the catch-you-off-guard element that injects a bit of interview stress. The interviewer is also probing your ability and technique to deal with workplace realities that may have constrained your original goals.

Tell me about a problem situation where it was difficult for you to remain objective. Here is another catch-you-off-guard question. The interviewer is evaluating if you can distinguish between your feelings about the solution and the solution that is best for your organization.

If it appears that interviews require a lot of work, you made a correct observation. This author conducted many interviews when administering an engineering recruiting program years ago while operating in the old benevolent, paternalistic work environment. Interviews then were more casual and easier for both sides. Today interviews are more intense, reflecting the faster pace of our globally competitive economy. The need for more rigorous interview-skill development today is driven by our new economy. View this need as one more competitive skill that you will enjoy continually refining throughout your career.

7

CAREER PLANNING

Goals and Your Personal Vision

CAREER PLANNING

In the preface and introduction, we discussed that, in years past, career planning was the joint responsibility of you and your employer. In today's employment culture, however, career planning is increasingly your responsibility. Today's fast-paced global economy forces business leaders to assume a short-term financial focus, making long-term planning a luxury. Part of this perspective is that today's employees are viewed as short-term assets, and career planning for the employee becomes an economically unjustifiable luxury. The net result is that long-term career planning is your job. Why does this arrangement prevail?

A SHORT HISTORY OF EMPLOYMENT ARRANGEMENTS

Engineers work in industry, government, colleges, and engineering firms ranging in size from a few people to thousands of employees. But to understand the recent evolution of employment arrangements as well as the future of the workforce, we need to focus on what has happened in industry. Most of the innovations in workforce arrangements originate in academia, more specifically the MBA programs, but are initially applied in industry.

These theories and arrangements are then migrated to the other economic sectors that employ engineers.

Before the Industrial Revolution, careers were dictated by family, gender tradition, and socioeconomic status. For most men, their careers were the same as their fathers and brothers. Sons of farmers became farmers, sons of coal miners became coal miners, and so on. Even when the younger generations decided to leave farming, mining, and other such careers for production jobs, the family tradition of sons repeating their fathers' occupations continued. Women had more limited career choice options, dictated by convention and social mores. Women could be homemakers, nurses, or teachers. For both men and women, career progression and career ladders were, with few exceptions, nonexistent. World War II changed all this.

Prior to World War II, a few large companies dominated their markets and the U.S. economy. Following the war, the economy rapidly expanded, resulting in the start-up of many more companies. The corporate organization became the driving force in American business. The corporation provided opportunities for employment, career growth, and economic well-being for both men and women. It was in the best interest of the corporation to train workers and provide lifetime employment after they were trained. A culture of "entitlement" was created based on an implied contract. In return for the employee's loyalty, employment until retirement was assumed. Career planning was unsophisticated. Male workers expected periodic promotions until they reached a plateau where they would remain until retirement.

During World War II, because of the massive departure of males from the workforce to the military, corporations were forced to employ women in production positions, as illustrated by the Rosie the Riveter publicity campaign. But after the war, women again had limited career options and limited career mobility. Within the organization in the '50s, '60s, and '70s, many women were still found in secretarial, administrative, and clerical roles. Both men and women also found that when they climbed the career ladder, they had to live with their decision because there was little internal mobility, little mobility from one company to another, and little mobility from one occupation to another. That all changed in the last quarter of the 20th century.

The late 1970s saw dramatic shifts in the labor force. Women entered the workforce in greater numbers in a much broader range of positions and occupations. There were massive numbers of layoffs as companies struggled to remain in business. The high-tech industry emerged. There was a breakdown of the implied social contract between employee and employer (loyalty to the employer in return for job security). All these changes resulted from multiple influences, including the Vietnam War, changing social conditions, deteriorating economic conditions, the popularization of the computer, the feminist movement, the baby boom, and so on.

We have now migrated to the Age of Multiple Choices. The complete integration of the Internet into our socioeconomic structure has created yet another set of influences. We are now in a global economy. It is now easy to create and operate global design teams. It is now much easier to manage employees located around the world. Business competition and your competition in the job market has gone global, and it seems as though everything in the world economy moves faster. As an employed engineer you now have a variety of career options or career ladders to select from. You can continue to stay with your employer and be considered for promotions. You can more freely change employers to advance your career. Or you can choose from a variety of self-employment options. On the other hand, the employers of engineers can continue to have engineers on their payroll, they can hire engineers as consultants or job shoppers, or they can outsource all engineering functions.

A SHORT PRIMER ON CAREER PLANNING

With all these choices, your need for careful individual career planning becomes critical. Assume that you are an independent economic unit, and as such, the ultimate responsibility for your career plan is your responsibility. With today's rapidly changing technology and work environment, you should think in terms of "serial careers." Throughout your working life as an engineer, you will be faced with more opportunities, more alternatives, more career decisions, and more career crossroads. Just as organizations change to fit the times, so too must you adjust and adapt to the organizational changes that will occur.

The steps in career planning are the following:

1. Define your short-term and long-term career goals.
2. Review the mission and goals of your employer.
3. Evaluate the compatibility of your goals with your organization's strategic or business plan.
4. Identify your strengths and weaknesses. Perform a self-assessment of your skills, abilities, motivations, interests, values, temperaments,

experiences, and accomplishments. The intent is to develop a firm foundation of information about yourself and then be able to compare that foundation to the requirements of other positions within the organization. Use a formal test if one is available at a reasonable cost. A free self-assessment questionnaire and evaluation is available at: *http://academic.engr.arizona.edu/vjohnson/PersonalCareerAssessment Questionnaire/PersonalCareerAssessmentQuestionnaire.asp.*

5. Identify the next two or three positions which you would like to attain. Determine the requirements of these positions.

6. With the help of your supervisor or a mentor, evaluate the development you need to meet these requirements.

7. With the help of your supervisor or a mentor, design a program to attain the necessary knowledge and skills. Within this program you need to establish priorities—when do you want to accomplish each goal? Within this program, you also need to identify barriers and how you will overcome them.

8. Implement the program.

9. Periodically assess your progress; restart this process as necessary.

Note that in the current Age of Multiple Choices, your career path may cross over multiple functions and multiple companies. While most career paths are perceived as an upward path through one functional organization, you should not feel constrained from pursuing the multitude of possibilities that exist within a cross-functional career path. This can be a part of your plan or can be a "plan B" alternative.

Also note that your multiple choices may include the nontraditional. As organizations have experienced re-engineering, downsizing, and other changes, the result has been flatter organizations (this process is sometimes called "vertical disintegration") that provide less opportunity for career advancement through promotion. For example, when I started in industry, there was the following engineering hierarchy for promotions: assistant engineer, engineer A, engineer B, senior engineer, consulting engineer, supervising engineer, engineering department manager, and engineering vice president. Other organizations had many variations of promotional sequences and titles. Now, instead of multiple engineering levels, jobs are redesigned so as to continue to challenge employees to do their best work without changing levels or perhaps without changing titles. A typical result would be a reduction of eight levels, as just described, to only four or five levels.

How is your career path defined when the number of levels or title changes is reduced? There are three ways to do this:

1. *Job enlargement.* Broadening the scope of a job by expanding the number of different tasks to be performed
2. *Job enrichment.* Increasing the depth of a job by adding employee responsibility for planning, organizing, controlling, and evaluating it
3. *Job rotation.* Shifting a person laterally from job to job

Each of these strategies can allow the engineer to learn new skills and to further refine and develop existing skills, thus preparing for advancement opportunities when they do occur.

WORKPLACE OF THE FUTURE

To plan for the future, you may be wondering what will happen to current workplace arrangements. What can you expect of the future workforce? How can your goals be compatible with future workforce directions? There are numerous references about the future workforce. One in particular is *The World is Flat* (Thomas L. Friedman; Farrar, Straus, and Giroux; New York, 2005).

The following material is gleaned from a Rand Corporation report that was prepared for the U.S. Department of Labor (Lynn A. Karoly and Constantun W. A. Panis, "The 21st Century at Work," Santa Monica, California, 2004).

Three major forces will influence the U.S. workforce in the 21st century:

1. Demographic trends predict a workforce that will not grow as rapidly as in the past, and skill will become more a defining characteristic of workers. The slowdown in growth is attributable to a slowdown in population growth, to an aging population, and to numerous other factors.
2. Ongoing technological progress will continue to reshape multiple aspects of the workforce.
3. There will be increased globalization of the U.S. economy enabled by advances in information technology and communication technology as well as other factors.

The workforce changes driven by these three forces will in turn cause changes in the following areas:

- The organization of production will result in vertical disintegration (flatter organizational structure) and more specialization, and acquiring and sustaining knowledge will become more important as a means of attaining competitive advantage. This implies that as you plan your career, you should seriously consider becoming a technological specialist.
- The nature of the employment relationship and employment location will change. More work will be performed in alternate arrangements such as self-employment, contract work, and temporary help. More work in firms will be done by self-employed individuals, sometimes called "e-lancers." This implies that your career plans should include procurement of a professional engineering license (see Chapter 17 on the licensure process) and openness to self-employment (see Chapter 15 on moonlighting).
- Concerns for safety and security will increase concurrently with concern for preservation of privacy. This is due to our shrinking world and global economy. This issue can impact your career plans and can provide more career specialization opportunities.
- Future technology developments will require all workers, especially engineers, to be more adaptable throughout their career to changing technology and product demand. A wonderful book that explores future technologies and their societal impacts is *Engineering Tomorrow*, a compendium of articles by experts edited by Janie Fouke, Trudy E. Bell, and Dave Dooling. (IEEE Press, Piscataway, New Jersey, 2000).
- Demand will increase for employees who are adaptable to changing technology and product demand. It is important that your career plan includes contingencies for these factors.
- There will be continuing changes in the wage and benefits arrangements. The mechanisms driving wage disparities, or the gaps between upper and lower wage earners, will continue. Demand for skilled workers will increase—good news for engineers. However, access to fringe benefits from employers will decrease. Therefore, your career and economic plans need to consider this fact.

In the future, we will see the paradigm of knowledge-based organizations, where intellectual capital becomes an important asset for generating competitive advantage. Engineers can be a key ingredient of this intellectual capital.

We will also see large companies divided into semiautonomous or autonomous units that interact as though they are separate companies. Global teams will be assembled for specific projects and subsequently dissolved. There will be rapid prototyping using computer-aided design and rapid fabrication methods in manufacturing. This more agile manufacturing capability will allow companies to outsource the production of goods designed and tested in-house. This forecasts an increase in global engineering teams for some organizations and the retention of in-house engineering in other companies. You need to monitor continually the direction of the workforce issues in your area of specialization and to keep your career plan flexible. You need to understand your technology specialty, markets, customers, suppliers, business processes, and other invisible assets of your organization.

All this may sound complicated, and you may feel in need of a consultant. If your organization has a human resources department, it can provide information and assistance. Alternatively, search the Internet or your local library. Some engineering association Web sites have information about career planning and the directions of the future workforce.

YOUR GOALS

People with goals succeed because they know where they're going.
Earl Nightingale

The first step of career planning is to establish goals. If you ever participated in team sports, you learned the value of team goals. If you were involved in individual sports, the value of personal goals soon became very clear to you. For your career, you also need to set goals and regularly review them.

As an engineer, implementation of your goals is part of the learning process that must continue for the rest of your life. You need to learn new technologies and new skills to help you progress in your career. The following six-step process will help you manage your goals:

1. *State the goal.* Identify a significant purpose that you want to achieve.
2. *Define the objective.* Outline what steps you must take to achieve that goal. For example, you and your supervisor agree that you should improve your skills in oral presentations or learn a new computer program.
3. *Outcomes.* Define the specific knowledge, skills, and attitudes you'll need to develop to accomplish the objectives that you have listed. Determine what you can measure to know you have attained the goal.
4. *Learning activities.* Identify and participate in appropriate learning activities to achieve the desired outcomes, including courses and/or experience.
5. *Assess performance.* Continue learning activities until you can perform the task and achieve the goal.
6. *Repeat the cycle for more goals as needed for your career plan.*

Note that these steps look the same as the steps above for planning your career. They are very similar but are applied at a macro level for career planning and at a micro level for goals.

A person should set his goals as early as he can and devote all his energy and talent to getting there. With enough effort, he may achieve it. Or he may find something that is even more rewarding. But in the end, no matter what the outcome, he will know he has been alive.

Walt Disney

YOUR VISION

If you don't have a vision, nothing happens.

Christopher Reeve

Goals may be too variable and may be incompatible with one another if they are not tested against your personal vision. Do you have one? Did you think goals are only for CEOs because effective CEOs seem to be the only ones that regularly articulate their vision?

What is personal vision? The CEO's vision defines what the executive wants the organization to be. Your personal vision is what you want your life to be. It is a direction or guiding light for your life. It is a description of how you want to live your life. It defines the fundamental core values, passion, or reasons that you chose to live your life in the fashion you defined. A personal vision helps you define the types of goals that take you in the proper direction. It will still stand when all of the goals or tasks you defined have been completed and always will be a framework for setting new goals.

To develop your personal vision, concentrate on the things that are important to you. Categorize them in a list of areas of your life: family, extended family, career, professional, moral and ethical standards, physical and psychological health, hobbies and recreation, community, politics, financial security, faith, and other considerations that may be important to you. Yes, you are an engineer, but for this exercise, you need to be an amateur philosopher. Use your engineering talent to build a grid. For each of these life areas, define your vision, your goals to attain that vision, and tasks to achieve that goal. Do it in pencil, because these vision elements may change over time.

It is interesting to note that the results of this personal vision exercise can also be used to create your "ethical will." This is an old but interesting concept, not related to career planning, that leaves a clear record for your family and heirs of who you were, what you believed, what you learned in life, and what you valued. You may become especially interested in this concept starting in midlife. More information is available at *www.ethicalwill.com*.

Before you developed your personal vision, you may have established numerous goals. Now test them against your vision. Also, test the mission and vision of your employing organization against your vision. If there are significant differences or large gaps, you need to evaluate them and determine what action may be needed in your career plan.

Having a well-defined personal vision is especially important if you aspire to being a leader at some point in your career. The same is true if you aspire to be a leader in any other area of your life. (Leadership is discussed in Chapter 9.)

Give to us clear vision that we may know where to stand and what to stand for.
Peter Marshall

When you have a sense of your own identity and a vision of where you want to go in your life, you then have the basis for reaching out to the world and going after your dreams for a better life.
Stedman Graham

C h a p t e r

8

ORGANIZATIONAL MISSION AND VISION STATEMENTS

To evaluate a potential employer or your current employer, we recommend that you study the mission statement, vision statement, and core values statement of the company. These can usually be found in the business plan and in the "About Our Company" section of the Web site. If the company is proud of these statements and feels strongly about promoting them, they will be ubiquitous. You may find them on the walls, on multiple brochures, or even on the back of business cards. Mission statements and vision statements can tell a lot about an organization, so it's important to take time to review them.

If they do not exist for an organization, their absence also tells you something. It indicates that the organization may be quite weak in future planning and direction.

WHAT IS A MISSION STATEMENT?

A *mission* is your organization's fundamental purpose. It is the reason why it exists as a company. You can also think of a mission statement as a cross between a slogan and an executive summary. An effective mission statement should be able to tell your company story and ideals in less than 30 seconds. Also, the organization and its employees must actually believe in the mission statement—if they don't, it's a lie, and the customers will soon realize it. A mission is essential to staying the course. Ebbs and flows in the economy—as well as lucrative business opportunities—can complicate the ability of an organization to stay true to its identity. A solid mission sounds like a battle cry, uniting all levels of an organization.

WHAT IS A VISION STATEMENT?

A *vision* is the organization's view for the future—where it wants to be and what it wants to become. The vision is a compass to help the company know where it is headed as an organization. It is the long-range goal, what the company is working toward. Often, the vision is driven by the chief executive officer. Vision is closely related to dreams. If leaders fail to dream big dreams, then a company is destined to never attain greatness. The Pygmalion effect—or self-fulfilling prophecy—reminds us that we will only rise to the level of expectations laid upon us. Organizations that lack vision are doomed to failure as the competition moves a step ahead.

WHAT ARE VALUES?

Values, sometimes called core values, are often more recognizable than a mission or vision. Great companies with great values put their money where their mouth is. For example, values yield results in the form of good employee benefit packages, positive community involvement, and strong corporate cultures. For engineering organizations, values could include quality and excellence, honoring promises and contracts, and active support of our professions. Values are often the most outward sign for both employees and community of what your company is really all about.

WHAT DO THESE MEAN TO YOUR CAREER?

Understanding and accepting the organization's mission and vision statements are very important. If, because of your personal plans and convictions, you cannot buy into even a part of your organization's mission and vision statements, you have a conflict that needs to be resolved.

Mission, vision, and values play an intricate role in the organization's corporate culture. Organizational culture, discussed in Chapter 1, includes both the people and the work environment. Organizational culture—as defined by the mission, vision, and values—can either positively or negatively affect the behavior and the career success of the individual.

ON THE JOB

9

LEADERSHIP

"People cannot be managed. Inventories can be managed, but people must be led."
H. Ross Perot

Let's assume that at some point in your engineering career you would like to become a manager of some sort, whether along the technical track (project or department manager) or along the management track. It is very important for you to understand that management and leadership are distinctly different concepts.

Management is the practice of making things happen according to generally accepted rules and policies within a firm. It is working toward producing a design or a set of plans for the construction or manufacturing of a product. It is the task of managing people, technology, and equipment to obtain the objective.

Leadership is the ability to motivate and inspire a group to achieve collective and individual goals. Leadership skills relate to motivating people.

There is no better time to think about leadership skills than at this very moment. Although it is common to hear that there are "born leaders," the truth is that the most effective road to leadership is to study its principles and to practice them—all the time. Effective leaders never stop pursuing improvement of their leadership skills. In fact, there is an entire industry of professionals writing books on leadership, conducting seminars, and even running multisession training courses on leadership. Some of the organizations that hire young engineers will have in-house leadership training programs. Leadership is a recognized asset throughout the business world.

What do we mean when we talk about leadership? The initial response takes up only a few lines:

- A leader has vision.
- A leader understands himself or herself and others.
- A leader can effectively communicate his or her vision.
- A leader can inspire people to work toward implementation of the vision.

What is vision? In the context of leadership, it can be a goal for the future or a beacon on the horizon toward which an organization will navigate. It may include a single goal or several. It is a place removed from where the organization is at the moment. While I can't know for sure, I can imagine some visions from well-known people:

- The Wright Brothers' vision may have been for humanity to have the ability to move from place to place by flying in some sort of machine.
- For Bill Gates, it may have been for everyone to have ready access to electronic devices that would be able to solve complex problems with incredible speed.
- For Susan B. Anthony, it was for women in the United States to attain full voting rights.
- For Martin Luther King, it may have been to reach a time when persons of all racial heritages would stand together in the world enjoying equal rights. Reverend King's vision is well-known—"I have a dream."

A vision is virtually useless unless the visionary can mobilize forces toward attainment of the goal. The persons listed above certainly were able to do that. They all had the leadership traits necessary to make that happen. Like them, a leader must possess the ability to communicate the vision to those who are in a position to pursue it. A leader must be a good communicator. But good communication is not the only necessary skill. The leader must be able to inspire others, must exhibit a passion for the goal, and must be able to explain the value of the objective convincingly. The organization must perceive value in the vision—must buy in—for the program or initiative to be successful.

What skills and traits should a leader possess to inspire people to follow his or her vision? We could create a list of enormous length, but here are some of the more important ones:

- Integrity/trustworthiness
- Accessibility
- Good communication skills
- Understanding of self
- Understanding of others
- Good motivational skills
- Fairness
- Kindness
- Passion
- Cheerleader/champion
- Patience
- Reliability/consistency
- Unselfishness/team player
- Visibility
- Optimism
- Enjoyment of the organization

As you can see, the list of essentials is long, and it could be a lot longer. The list is idealistic, in the sense that many leaders possess some, but not all, of the items. Nevertheless, these skills and traits, many of which may come naturally to you, will contribute to your growth as a leader. You may wonder if you are interested in being a leader, because your image of the leader in your organization may be the CEO, the president of the corporation, or the chief engineer or it may be the mayor or one of many visibly prominent positions. Rest assured, those folks ought to have good leadership skills, although many do not. As a young engineer at the start of a career, you may feel totally remote from any role requiring leadership. This is not true.

Leadership is a valuable skill at any level. If you work with others, you may not be their supervisor, but you can be a leader. We often hear of "leadership in the locker room" in sports. What is that all about? It is about team members who have the ability to inspire others at their same level to perform, to chase a vision (World Series, Super Bowl, Ryder's Cup, etc.). It is about motivating teammates to perform at the peak of their ability. It is about success at what a team is about. Surely the head coach is the recognized leader of the team, at least officially. A head coach, without leadership skills, can drag a team down. But even a head coach who is a master leader needs leaders in the locker room. If you don't think so, just listen to the press conferences of the coaches, especially those of winning programs.

By now, you are wondering how to become a good leader. It is a process. The first thing you should do is to create a personal vision for yourself. What do you want to do with your life? What do you want to be in 10 or 20 or 30 years? Then, once you have that in mind, you need to inspire yourself to go for it.

One of the ways to begin leading others is to pay attention to yourself, to understand your motivational system, and to pay attention to those around you whose motivational systems or personality types may not coincide with yours. Many books have been written and seminars offered on the subject.

You may also share leadership issues with your mentor, should you have one. If not, you should consider finding one. (See Chapter 20 for more information on mentoring.) If you are in an organization that encourages personal and professional development, let it be known that you would appreciate any opportunity that might come along with respect to leadership training. It is likely to bring positive attention to you as an employee if management learns that you have an interest in becoming a good leader.

As a secondary effect, the leadership skills you learn in your role as an engineer won't be confined to your performance on the job. The skills spill over into your personal life. You will become a better family person. Your social and spiritual life will also benefit. If you want to experiment with leadership styles, you can do it pretty safely in those roles.

In closing, you are urged to participate in at least one professional association. We have found that participation in professional and technical societies furnishes excellent opportunities for leadership training and for trying different leadership styles. An excellent resource for enhancing the communication skills required for good leadership is participation in Toastmasters International (*www.toastmasters.org*). Professional societies offer many opportunities for leadership roles that can often be assumed at an early age. The risk of trying different leadership styles at an engineering society meeting is minimal. No one gets fired or demoted in that environment. It is sort of like going to a health club to work on physical development, only in this circumstance you will be working on personal development as a leader. I believe my leadership style has benefited greatly from volunteer work.

10

PROFESSIONALISM AND ITS IMPORTANCE

WHAT IS "PROFESSIONALISM"?

We all know that engineering is a "profession," that those who practice it may become "professional" engineers, and that we are expected to display "professionalism" in our work. What are those words really expressing? Actually, the concepts are widely recognized but may have different meanings to different people. We talk about these terms in this section, hoping to give you a solid foundation for building and expanding your own philosophy as your career matures. Let's begin by looking at several published definitions that are applicable to us.

"A calling requiring specialized knowledge and often long and intensive academic training"
Merriam Webster Online Dictionary

"A vocation or occupation requiring special, usually advanced education and skill (e.g., law or medical profession).
Also refers to whole body of such profession. The labor and skill required in a profession is predominantly mental, rather than physical or manual."
Black's Law Dictionary, fifth edition

*"An occupation such as law, medicine, or engineering that requires
considerable training and specialized study"*
American Heritage Online Dictionaries

*"A profession is an occupation that requires extensive training
and the study and mastery of specialized knowledge and usually has a
professional association, ethical code, and process of certifying or
licensing. Examples are accounting, law, medicine, finance, the military,
the clergy, and engineering."*
Wikipedia

The term *profession*, as we shall apply it, is not exclusive. For instance, it can
also be applied to those who receive payment for the performance of some sort
of service or activity. A well-known example is the professional athlete who is
compensated for playing a sport that amateurs play for the sheer joy of it. Not
many people would confuse the occupational duties of a professional engineer, doctor, or lawyer with those of an NFL linebacker, however. We subscribe
to the premise that a professional is not only one who has attained special
knowledge and skill through advanced training and study but that the occupational efforts of a professional are applied to the advancement and betterment
of humanity. That definition is widely recognized and demonstrated by licensure laws throughout the nation, as well as in codes of ethics of various professional societies and associations. The laws and codes charge us with the
protection of the health, safety, and welfare of the public through our professional activities. That responsibility is of paramount importance, taking prece-
dence over any other duties that may come into conflict with that lofty goal.

In Chapter 11, we will introduce you to a pair of solemn oaths taken by
engineers under certain circumstances. To underscore the fact that a heavy
emphasis is placed upon an engineer's duty to protect the safety, health, and
welfare of the public, we are going to take excerpts from each and repeat them
here.

EXCERPT FROM "OBLIGATION OF AN ENGINEER"

AS AN ENGINEER, I PLEDGE TO PRACTICE INTEGRITY AND FAIR DEALING,
TOLERANCE AND RESPECT AND TO UPHOLD DEVOTION TO THE
STANDARDS AND THE DIGNITY OF MY PROFESSION, CONSCIOUS ALWAYS
THAT MY SKILL CARRIES WITH IT THE OBLIGATION TO SERVE HUMANITY
BY MAKING THE BEST USE OF EARTH'S PRECIOUS WEALTH.

EXCERPT FROM "THE ENGINEER'S CREED"

AS A PROFESSIONAL ENGINEER,
I DEDICATE MY PROFESSIONAL KNOWLEDGE AND SKILL
TO THE ADVANCEMENT AND BETTERMENT OF HUMAN WELFARE.

I PLEDGE:
TO PLACE THE PUBLIC WELFARE ABOVE ALL OTHER CONSIDERATIONS.
IN HUMILITY AND WITH THE NEED FOR DIVINE GUIDANCE,
I MAKE THIS PLEDGE.

You can see that engineers who take these oaths are seriously committed to service to humanity and to protecting the health, safety, and welfare of the public.

Some professionals can be self-regulated (i.e., a designated professional organization will be authorized by a state legislature to promulgate rules and regulations and will be empowered to enforce them). For example, the U.S. legal profession is self-regulated by the American Bar Association (ABA). The ABA has the authority to promulgate rules of conduct for lawyers and to enforce them. With the exception of three states (Delaware, Florida, and Oregon,) the engineering profession in the United States does not have such an arrangement with respect to professional engineers. All the other states have legislation that places the authority in a state registration board under the oversight of a state agency.

What characteristics do members of the engineering profession need to demonstrate to attain the status of professionals? The lists below will go a long way toward answering that question.

Mandatory Characteristics of the Engineering Professional

- *Education.* Graduation from an Accreditation Board for Engineering and Technology (ABET) accredited engineering school with a bachelor's degree, or higher degree, in engineering
- *Participation in continuing professional competency courses.* As of this writing, continuing education is mandatory in 30 states; such courses must be acceptable to the state's registration board.

- *Demonstration of having mastered basic scientific and engineering knowledge and skills.* Evidenced by passing the Engineer in Training (EIT) or Fundamentals of Engineering (FE) examination of the National Council of Examiners of Engineers and Surveyors (NCEES)
- *Demonstration of having mastered a baseline level of engineering knowledge and skills.* Complete a minimum of four years of actual engineering practice acceptable to the registration board
- *Passing the NCEES practice examination in the discipline of choice.* Take this test subsequently to the required engineering experience
- *Holding a state-issued professional engineering license.* Hold a license in every state in which you practice
- *Adherence to state laws and regulations relevant to the practice of engineering*

Voluntary Characteristics of the Engineering Professional

The following are not required but are strong indicators of professionalism.

- *Participation in a professional association*
- *Recognition and adherence to a recognized code of ethics promulgated by a national engineering society.* See the appendix for the NSPE Code of Ethics and the NCEES Rules of Conduct.
- *Community involvement.* Preferably with an opportunity to provide input based upon sound engineering background and experience
- *Advancement and promotion of the engineering profession.* This occurs through various mechanisms, including but not limited to:
  - mentoring;
  - participation in Engineers' Week activities each February during Presidents' Week;
  - coaching the MathCounts competition, Future Cities Competition, Science Fair, or other engineering-based competition; and
  - participating in pertinent legislative and rule-making processes as an advocate of the engineering profession.

If you conform to all the mandatory items and participate in at least some of the voluntary ones, then you can truly call yourself a professional. As you read this, you may be asking yourself why you can't call yourself a professional without becoming licensed. The answer is that until you are a licensed professional engineer, the law does not consider you a professional. Your peers, your co-workers, or your employer may call you an engineer, but in the eyes of the law you are not entitled to use the title *engineer* until you have attained a license. Look at the Wikipedia definition at the beginning of the chapter. It says that a profession usually has a process of certifying or licensing. For engineering, it is not "usually"; it is always.

Chapter 17 deals with licensure and reasons to make that a goal. We have a strong and unswerving belief in licensure as an integral part of a commitment to the engineering profession. We believe it to be the cornerstone of professionalism in engineering. We urge you to make it a personal goal.

11

ETHICS FOR ENGINEERS

INTRODUCTION TO ETHICS

Simply stated, *ethics* is the study of moral conduct. It is, in the most basic terms, the ability to discern right from wrong. It is the ability to recognize moral duties and values. Ethical conduct, then, simply means doing right things and avoiding doing wrong or immoral things. So why does society spend so much intellectual energy and valuable time creating and promoting observance of lists of behavior known as "Codes of Ethics"? Are those lists, generally created by associations or societies, simply a flashy way of calling attention to the goodness of membership or of posturing for the benefit of the public? Are those codes designed just to attract members who can bask in the light of the organizational mantra?

Certainly you know the answer. A code of ethics is not a sham, not a glitzy show of membership value, not merely a collection of sweet-sounding words and phrases. Indeed not. In this chapter, you will learn something about the history of ethics, the many sources of engineering ethical codes, and the way that those codes weave their way into law.

Newspapers, radio, TV, and Internet news articles are crammed full of stories about ethical misconduct at all levels of our society. We hear of transgressions by individuals, corporations, organizations, and even government agencies. Codes of ethics, in and of themselves, will not harness such activity. For mainstream America, however, codes of ethics do provide guidelines for appropriate moral conduct. The way it works is that most people who respect themselves, who aim to behave within the confines of ethical behavior, will use the codes to help them navigate the various shoals of personal, business, and professional activity to examine their conduct against benchmarks of behavior.

The word *ethics* is derived from ancient Greek. *Ethos* means "the place of living." From *ethos* is derived the word *ethikos,* meaning the theory of living. Ethical conduct was an area of philosophical study many centuries B.C. The early Greek philosophers, such as Socrates (470–399 BCE), Plato (427–347 BCE), and Aristotle (384–322 BCE), led the way. They delved into the meaning of human instincts for moral living. Teachings in the Old and New Testaments focus on morality and ethical conduct; one of the most well-known codes of ethics is the Ten Commandments. The history of the world's major religions spans millennia.

Many philosophers believed that ancient humans, and perhaps even animals, had instinctual drives to behave morally within the context of their tribal environments. In simpler times, humans were preoccupied with hunting for food and in maintaining tribal harmony for survival. We are long gone from those days. Nevertheless, the need for ethical conduct is strong if we are to go about our life in relative harmony with our fellow citizens, our families, and our colleagues.

CODES OF ETHICS

Codes of ethics have been drafted and adopted by a multitude of engineering professional and technical associations. Codes generated by such organizations do not carry the force of law. Members of the organizations agree, when they sign on, to abide by those codes. If a member is shown to have violated any of the canons of ethics of the association, the association may, after a formal hearing process, choose to censure, suspend, or terminate the offender from membership. While such an experience can be embarrassing to the member, the association has no authority to fine or incarcerate the offender.

Codes of ethics for engineering societies do not have lengthy histories spanning back to the time of Aristotle, Plato, and Socrates. The search for a suitable code of ethics for engineers has a much shorter history.

In 1932, the Engineers' Council for Professional Development (ECPD) was formed. It was created by what is known as the Five Founder Societies, those being The American Society of Civil Engineers (ASCE), the American Society of Mechanical Engineers (ASME), The Institute of Electrical And Electronics Engineers (IEEE), the American Institute of Chemical Engineers (AIChE), and the American Institute of Mining, Metallurgical, and Petroleum Engineers (AIME). ECPD was formed to promote the status of the engineering profession and enhance the quality of engineering education. ECPD no longer exists, having been replaced in 1980 by the Accreditation Board for Engineering and Technology (ABET), which focuses its efforts on the accreditation of educational programs.

When ECPD was formed, it created a joint committee of national engineering societies and associations to strive for the development of a single code of ethics to which all the organizations could subscribe. Although the goal was a noble one, it has never been achieved. The development of engineering society and association codes of ethics has, however, continued through today.

Society and association codes of ethics are not static, being amended or upgraded from time to time to take modern developments into consideration. For instance, codes of ethics recently have been upgraded by various professional and technical societies to reflect environmental concerns as ethical responsibilities, and some codes were upgraded to incorporate sustainable development as an ethical objective. While the codes of ethics of various engineering societies are not identical, they are quite similar.

During the early years of development of engineering codes of ethics, a sentiment was held throughout the engineering community that a widely accepted, short form of a philosophy could capture the essence of the duty of an engineer. In 1954, the National Society of Professional Engineers (NSPE) adopted the Engineer's Creed, authored by Paul H. Robbins, PE, then NSPE's executive director. The Engineer's Creed has survived until this time and has become widely used in various ceremonies, including installations of society or association officers, graduation ceremonies, presentations of licenses to newly licensed engineers, and induction into various engineering groups.

THE ENGINEER'S CREED

AS A PROFESSIONAL ENGINEER,
I DEDICATE MY PROFESSIONAL KNOWLEDGE AND SKILL
TO THE ADVANCEMENT AND BETTERMENT OF HUMAN WELFARE.

I PLEDGE:

To give the utmost of performance;

To participate in none but honest enterprise;

To live and work according to the laws of man

and the highest standards of professional conduct;

To place service before profit, the honor and

standing of the profession before personal advantage,

and the public welfare above all other considerations.

In humility and with the need for divine guidance,

I make this pledge.

Engineers take a similar pledge called the "Obligation of an Engineer" as they are inducted into the Order of the Engineer. The Order of the Engineer is an organization that originated in Canada in 1926 and was started in the United States in 1970 to foster a spirit of pride and responsibility in the engineering profession. We strongly recommend that you seek out the Order of the Engineer and join for the uplifting experience (*www.order-of-the-engineer.org*). There are no further meetings or duties associated with affiliation. Those of us who have experienced the ceremony find it to be highly motivational toward ethical conduct and pride in our profession, and we wear a simple stainless steel ring on the little finger of our right hands, providing a readily recognized symbol to others who have taken the oath below

OBLIGATION OF AN ENGINEER

I am an Engineer, in my profession I take deep pride.

To it I owe solemn obligations.

Since the Stone Age, human progress has been spurred by the engineering genius. Engineers have made usable Nature's vast resources of material and energy for Humanity's benefit. Engineers have vitalized and turned to practical use the principles of science and the means of technology. Were it not for this heritage of accumulated experience, my efforts would be feeble.

As an Engineer, I pledge to practice integrity and fair dealing, tolerance and respect, and to uphold devotion to the

STANDARDS AND THE DIGNITY OF MY PROFESSION, CONSCIOUS ALWAYS
THAT MY SKILL CARRIES WITH IT THE OBLIGATION TO SERVE HUMANITY
BY MAKING THE BEST USE OF EARTH'S PRECIOUS WEALTH.

AS AN ENGINEER, I SHALL PARTICIPATE IN NONE BUT HONEST ENTERPRISES.
WHEN NEEDED, MY SKILL AND KNOWLEDGE SHALL BE GIVEN WITHOUT
RESERVATION FOR THE PUBLIC GOOD. IN THE PERFORMANCE OF DUTY
AND IN FIDELITY TO MY PROFESSION, I SHALL GIVE THE UTMOST.

As members of the Order of the Engineer, we are still in awe of the duties imposed upon us by the oath, and we take them quite seriously. We can honestly and earnestly commend you to not only become a licensed professional engineer but to seek out and join the Order of the Engineer.

CODES OF ETHICS IN THE LAW

Interestingly, the codes of ethics of various professional organizations have found their way into the law through the rule-making process. Rules are drafted and promulgated by professional registration boards in all the U.S. states and territories. There are many regulated professions. Some of the regulated classes include lawyers, physicians, nurses, architects, land surveyors, landscape architects, and, of course, professional engineers. When one examines the regulations of the professions, particularly with respect to lists of actions that are considered misconduct, one will find a significant relationship between legally prohibited behavior and the codes of conduct of the professional organizations. Needless to say, violation of the regulatory codes of conduct can result in a hearing and, if one is found guilty, loss of license, a monetary fine, and even jail time.

It is not altogether surprising that associations' codes of ethics and regulations promulgated by boards of registration are similar. The registration boards in the United States and its territories are members of the National Council of Examiners of Engineers and Surveyors (NCEES). NCEES has developed language for a model licensure law and for model rules. Although the member registration boards are not obligated to adopt the model rules, the NCEES models are closely paralleled in many jurisdictions. The NCEES model rules are similar to the codes of ethics of the major engineering organizations. It is not surprising, then, that the association codes, although not bearing force of law, will closely parallel the regulatory constraints of ethical conduct for engineers.

The NCEES Model Rules of Professional Conduct are included in the appendix for your study and future reference. The NCEES Model Rules of Professional Conduct set down three classes to whom a professional engineer has ethical duties. They are:

- society;
- one's employer and clients; and
- other licensees.

To know the extent of adoption by any particular jurisdiction, one will need to obtain the applicable regulatory rules and review them. Those rules should be near the top of the required reading list of every graduate engineer, including professional engineers and those aspiring to become professional engineers. Remember, an alleged violation of any of the ethical constraints in a state's regulations can lead to a hearing before the registration board, resulting in a significant penalty if the allegation is proven. Loss of license is a serious setback and can result in one's inability to continue practicing engineering. Even for engineering graduates who choose to rely upon the industrial or, where applicable, government exemption, ethical conduct should be a matter of personal pride and commitment. The published codes will give one an excellent frame of reference for specific ethical issues peculiar to engineering.

PROFESSIONAL LIABILITY ASPECTS OF ETHICAL CODES

Chapter 12 discusses professional liability. There you will find that proving negligence against an engineer requires the party making the claim to prove certain elements. One of them is that the engineer owed a duty to the party and then breached that duty.

One of the sources of engineers' duties is ethical and professional responsibility. They are not contractual duties but arise from the obligations of licensure. As a licensed professional engineer, one is duty-bound to practice legally and in accordance with the laws of the land.

Furthermore, the duty of a licensed professional engineer includes conformity to recognized ethical codes. As noted above, the ethical codes of engineering associations are paralleled in many of the regulatory codes of conduct, so the party making a claim against an engineer can examine a double set of rules. For engineers who do not choose the licensure route, there may be a lesser duty to comport with the licensure laws and regulations, but the codes of ethics likely will emerge in a case against even an unlicensed engineer.

It is incumbent upon every licensed professional engineer to become familiar with at least one major ethical code and to review it periodically against his daily conduct on the job. It can be argued that it is as important to stay in touch with ethical constraints as it is to keep up with technology (and it will take less time).

ETHICAL DILEMMAS

So far, we have discussed the easy stuff. In practice, there will be times when more than one ethical canon will come into play and the practicing engineer will have a dilemma. One of the most common types of situations occurs when an employer or a client is bending the rules or actually engaging in illegal conduct.

As a hypothetical example, assume that you are working as a project manager on a site plan for a land developer, and you or someone in your firm becomes aware that some contaminants were illegally dumped on-site in the recent past. The EPA lists the contaminants as extremely hazardous. You meet with the client to deliver the bad news. In fact, the news is extremely bad, because an extensive amount of environmental evaluation will be needed to decide how to clean up the mess and determine the method of disposal. It will probably eat up enormous sums of your client's funds, and slow the project's progress to a snail's pace because of the bureaucratic process of getting a cleanup and disposal plan designed, approved, and implemented. Your client will not only need to foot the bill, pending the unlikely recovery from an unknown dumper of its losses, but the client will have to wait an unpredictably long time before any other work can be done.

The client asks to meet again with your supervisor or even the CEO of your firm. At that meeting, the client and your superior both tell you to back off. This thing can be handled, and you will not have to worry about it. You are reminded that you, as well as your firm, have a duty to your client that requires you to act in its best interest. In fact, the NSPE code of ethics says that a professional engineer shall act for each employer or client as faithful agent or trustee. That puts a heavy burden on you to place your client first, doing nothing to harm the client. The code also says, however, that if an engineer's judgment is overruled under circumstances that endanger life or property, he or she shall notify the employer or client and such other authority as may be appropriate. In this example, you have already told the employer and the client. The only one left is "such other authority." Now that is scary. If you do that, you can hardly imagine that your job will be intact for very much longer.

Now, here is the clincher. The NSPE Code of Ethics says that a professional engineer will hold paramount the safety, health, and welfare of the public. Your state law and code of conduct also impose that duty upon you. *Paramount* is clearly the operative term. It means "above all else."

The situation described above is an awful one, but you can rest assured it has happened in real life. In the example, you are being pushed to the brink, and your choices are limited. You could go on, stay silent, and hope that no one gets ill or dies down the road because of whatever secret plan your client has hatched. Or you can quit your job, hoping to escape the reality. Of course, there is no guarantee that your name won't come up sometime down the road. You can also make the hard choice and do the ethically required thing: Report this issue to the authorities and become a "whistleblower."

There have been some well-documented failures in recent history when engineers' judgment was ignored or overruled. One of the most egregious was the loss of the space shuttle Challenger. On the prelaunch night of January 27, 1986, the temperature at the launch site was unusually low. Morton Thiokol engineer Roger Boisjoly sat in meetings with his superiors, pleading with them to have the launch aborted. Morton Thiokol was the designer of the space vehicle's solid rocket motor. The company's engineers had known for a very long time that the rubber O-rings designed to retain the rocket motor propellant were unreliable in cold temperature. The O-rings had been known to allow blow-by of superheated gases in such conditions. NASA had been previously informed but decided not to pursue redesign because of cost factors.

Boisjoly implored his supervisors, as well as NASA launch officials via teleconference, to abort. NASA was under various pressures to launch as scheduled. One of the pressures came from the public focus on the "Teacher in Space" program, which would send the first civilian astronaut, schoolteacher Christa McAuliffe, into space aboard Challenger. NASA, however, left the decision to Morton Thiokol, who eventually made the decision at the upper-management level to launch, even though Boisjoly and other engineers never abandoned their position that the launch would likely result in failure of the O-rings.

At 11:38 AM on January 28, 1986, Challenger left the launch pad. A short 73 seconds later, Challenger disintegrated, killing all seven of her crew, including Christa McAuliffe. The financial loss to NASA and the nation was enormous, outdone only by the needless loss of seven courageous, highly trained young astronauts.

Subsequently, Boisjoly and other Morton Thiokol engineers were summoned before the Presidential Commission on the Space Shuttle Challenger Accident. Boisjoly told the commission, in great detail, the events preceding the launch and the knowledge that the engineers and management of Morton Thiokol had that had predicted the tragedy. Morton Thiokol's CEO, Charles Lock, was quoted as saying, "People are paid to do productive work for our company, not to wander around the country gossiping." Boisjoly was not terminated, but, reportedly, his work environment became unbearable. Six months after testifying before the presidential commission, he left Morton Thiokol. He is reported to have suffered from posttraumatic stress syndrome. He became a spokesperson on behalf of engineering ethics, addressing students at universities and appearing before engineering associations.

This true story may give you reason to wonder if being ethical is too much of a challenge or if the consequences can be too harsh. I think the answer to both questions can be yes, but the alternative can present personal distress as well. Who would you rather be in the Challenger story: engineer Boisjoly or the manager who authorized the launch?

WHISTLEBLOWERS

A *whistleblower* is defined as one who reveals wrongdoing within an organization to the public or to those in positions of authority. Whistleblowing is demanded by the canon of ethics, which states:

> Licensees shall notify their employer or client and such other authority as may be appropriate when their professional judgment is overruled under circumstances where the life, health, property, or welfare of the public is endangered.

In the Challenger incident, engineer Boisjoly did his whistleblowing after the fact under a requirement to testify before a presidential commission. What about other times when an engineer is challenged to solve a problem, such as the one in the hypothetical situation introduced in this chapter? Isn't it likely that the engineer will be putting his or her job on the line, maybe affecting the ability to earn a living? Yes, that is a possibility. However, many states have recognized the pressures that those questions put upon people who have to choose between the public good and their own day-to-day economic survival.

Most states have some sort of statutory or common-law "whistleblower" or antiretaliation laws. Not every lawyer will know about such state or corresponding federal laws, especially laws outside the lawyer's own state. If you are interested in looking into what states have such laws, you can find out at WhistleblowerLaws.com (*www.whistleblowerlaws.com/law.htm*).

As a forensic engineer in the civil and land surveying sector in New Jersey, I was engaged as an expert witness in a case in which a licensed Professional Engineer was suing his former employer for wrongful discharge. He maintained that his discharge was retaliation resulting from his informing a local building code official of construction of a facility on an industrial site without benefit of a building permit. His firm had told him that its client's decision was to proceed and that it was none of his concern. Then his employer asked him to furnish resident engineering services at the site for the portion of the work being furnished by the company. When the engineer made inquiry at the municipal building department about the legality of this activity, the job was shut down until the owner obtained proper permits.

After reviewing a substantial amount of documentation produced in discovery, I reviewed his actions in the light of the New Jersey regulations concerning conduct by a professional engineer, as well as the NSPE Code of Ethics. Last, I reviewed the New Jersey Conscientious Employee Act. Putting it all together, I concluded that he had performed in accordance with his prescribed duties under all of those documents and prepared a report expressing my opinion to that extent. While I don't know the details, I was informed by counsel that his case was settled, resulting in compensation to him by his former employer, and that he was going on with his life. I was very proud of him and pleased to have furnished a report that helped resolve the issue.

NSPE CODE OF ETHICS

Because the NSPE Code of Ethics is one of the most commonly referenced codes, it will be provided for your future reference. You will find the NSPE Code of Ethics for Engineers in the appendix. It is suggested that when you read it, you pay attention to the similarities between it and the NCEES model regulations previously discussed. Your professional life will go a lot more smoothly if you take the time to read the code from time to time, reflect on it, and try to apply it to your daily tasks as an engineer.

Perhaps the most compelling portion of the NSPE Code is the opening preamble and list of fundamental canons.

Preamble

Engineering is an important and learned profession. As members of this profession, engineers are expected to exhibit the highest standards of honesty and integrity. Engineering has a direct and vital impact on the quality of life for all people. Accordingly, the services provided by engineers require honesty, impartiality, fairness, and equity and must be dedicated to the protection of the public health, safety, and welfare. Engineers must perform under a standard of professional behavior that requires adherence to the highest principles of ethical conduct.

I. Fundamental Canons

Engineers, in the fulfillment of their professional duties, shall:

1. Hold paramount the safety, health, and welfare of the public
2. Perform services only in areas of their competence
3. Issue public statements only in an objective and truthful manner
4. Act for each employer or client as faithful agents or trustees
5. Avoid deceptive acts
6. Conduct themselves honorably, responsibly, ethically, and lawfully so as to enhance the honor, reputation, and usefulness of the profession

The full Code then goes on to give detailed ethical guidance in each of the six areas.

Engineering is a noble profession, highly regarded by the general public. That is not to say that issues of ethical misconduct by some engineers do not arise in the news from time to time. It is my hope that neither my name nor yours will ever appear in such a context. It is my hope that this chapter has given you sufficient understanding of ethics and their value to all of us that you will not only remain a moral and honest individual, but that you will reach out from time to time to mentor colleagues and those who will follow in your footsteps in this very significant area of societal behavior.

12

PROFESSIONAL LIABILITY

There is a principle in the world of commercial products known as "strict liability." It is embodied in products laws in all the U.S. states and territories. The principle of strict liability for products applies to all parties in the chain of manufacture of any product that finds its way to a consumer. If a consumer is harmed by use of an inherently defective product, the manufacturer of the product and anyone who manufactured parts for the product as well as the vendor who sold the product to the consumer can be liable for the damages. Even if those parties were not negligent, or careless, or if they had absolutely no indication that the product was defective, the law of strict liability will hold them to be liable. This principle is not generally applicable to engineering projects, as explained below. The distinction, however, is important because professional liability for negligence or malpractice is viewed by the legal system in a special way.

For those engineers who offer their services to the public, primarily those in private consulting practices and in the construction field as well as in government practice, it's always possible that an aggrieved party will name them in a claim. A variety of claims can be made, but the primary one of concern is a claim of professional negligence resulting in damages to the claimant. You should have a solid foundation of understanding about professional negligence and about how you should conduct yourself as a practicing engineer.

SOME EXAMPLES

To clarify the nature of professional liability claims, I would like to present you with an example of one that was made against my firm over 20 years ago. A condominium association commissioned us to evaluate its roads and drainage structures. They were exhibiting signs of deterioration, leading the association to believe that the developer's construction was suspect. Within a few months, we prepared an extensive report detailing the condition of the roads based upon a numerical rating system published by the Asphalt Institute, and we identified dozens of storm drainage inlets that were structurally unsound. The board of directors accepted our report, and the state agency that monitors condominium developments praised it. The board entered into an agreement with the developer under which all the identified inlets and those roads that had ratings below certain levels would be repaired at the developer's expense. Our firm was designated to observe the work and determine whether it was acceptable. Remedial construction commenced, and all was proceeding satisfactorily.

While this was going on, an intense political contest was developing between the incumbent board and others who took issue with much of its activities, the aforementioned agreement being one of them. The election resulted in almost a complete turnover of the board's membership. The new board took office during the developer's remediation project, and things became tense almost immediately. The new board, feeling that its predecessor board had "sold out," attempted to have the developer do more work than specified in the agreement. The developer refused. The board asked us to reconsider our ratings of roads. We did not, because several years had passed and we could not return to the "original scene." In addition, all parties had accepted the report as valid at the time of completion. We explained the history of the project to the new board and made it clear that we were not party to the preparation of the agreement. The board was not satisfied, and it filed a claim against the firm for professional negligence.

Fortunately, our attorney called for a prediscovery meeting with the board and its attorney. It quickly became apparent at that meeting that the board really didn't have any specific material claims of negligence but was acting out its general hostility against the old board and the developer. The case was settled out of court. We forgave a small outstanding invoice and paid our attorney nearly $2,000 in legal fees, and the board released us from future claims. This example demonstrates that an engineering firm doesn't need to commit malpractice or be negligent for a claim to be made. Despite the fact that our firm suffered some financial loss as the result of the claim, the outcome would be considered acceptable in business terms.

Another major claim made against the firm ran over a period of nearly seven years between filing and trial. The outcome of the trial was that a jury found the firm to have not been negligent, and the plaintiff received no financial award from us. That's the good news. The bad news is that the seven years caused disruption of our normal work and emotional distress for the principals of the firm, and cost well in excess of $200,000 to defend, including about $50,000 of lost billable time.

Fortunately, our firm had very few claims made against it in nearly 30 years of practice. Why was that? It was because we made a point of practicing risk management and educating ourselves about professional liability. We would like to give you insights into this important area of knowledge early in your engineering career.

PROFESSIONAL NEGLIGENCE

First, we need to understand what constitutes professional negligence (or malpractice). *Black's Law Dictionary* defines *malpractice* as:

> Professional misconduct or unreasonable lack of skill. It is the failure of one rendering professional services to exercise that degree of skill and learning commonly applied under all the circumstances in the community by the average prudent reputable member of the profession with the result of injury, loss, or damage to the recipient or those entitled to rely upon them. It is any professional misconduct, unreasonable lack of skill or fidelity in professional or fiduciary duties, evil practice, or illegal or immoral conduct.

In virtually all legal jurisdictions in the United States, as design professionals, we generally are held to perform to the reasonable standard of care usually exercised by members of our profession in the same geographic area and at the same time as the alleged negligent act occurred. That doctrine is firmly enforced in courts of law. But as we will see later, it can be voided by faulty contracts or by offers to perform beyond the standard.

Once someone makes a claim of negligence against an engineer, that party must prove that the following four elements existed:

1. Duty owed
2. Breach of duty
3. Damages
4. Proximate cause

Let's discuss those elements.

Duty Owed

Duty owed by an engineer can appear in many forms, including:

- Contract between the engineer and the client
- Laws, ordinances, regulations, and codes (whether known to you or not)
- Ethical and professional responsibilities
- Common law

Breach of Duty

Your duties are not only to the parties to the contract but to the public at large, at least to the extent that it may experience a foreseeable injury resulting from your work product. Common causes of liability arising out of breach of duty include:

- Code violations
- Violations or failure to comply with laws, regulations, and permitting processes
- Unlicensed practice
- Errors and omissions
- Actions of subconsultants
- Failure to meet appropriate standard of care
- Breach of contract
- Breach of warranty or guarantee

It is important for you to know that privity of contract (the principle that a contract cannot confer rights or impose obligations on anyone except the parties to it) is not a requirement for a party to sue a design professional. The courts have long held that a third party who experiences an injury arising out of work performed by an engineer is entitled to seek compensatory damages from that designer, even in the absence of a contractual relationship. So, for instance, if a driver loses control of a vehicle and experiences damage to the vehicle, himself, or his passengers, he may seek monetary damages from all those who are perceived to be responsible, one of which is likely to be the engineer who designed the road.

Damages

Black's Law Dictionary defines damages, in part, as:

A pecuniary compensation or indemnity, which may be recovered in the courts by any person who has suffered loss, detriment, or injury, whether to his person, property, or rights, through the unlawful act or omission or negligence of another. A sum of money awarded to a person injured by the tort of another.

Compensation for damages will be made in money. In a claim against an engineer, the plaintiff will have to demonstrate that there has been harm that can be equated to a sum of money.

Proximate Cause

A partial definition of proximate cause taken from *Black's Law Dictionary* is:

The last negligent act contributory to an injury, without which such injury would not have resulted. The dominant, moving, or producing cause; the one that necessarily sets the other causes in operation.

This element is particularly important for the plaintiff to prove, for even if it can be indisputably shown that an engineer's design was flawed, that will not win an action unless the flaws can be proven to have been the cause of the plaintiff's damages. For example, if it can be shown that something you designed did not meet the legal or technical standard for sizing a storm drain pipe, your actions will not matter legally if the claim is for a physical injury sustained because a person fell from a balcony near the storm drain. The negligent act must be shown to have caused the damages.

Standard of Care

Now that all of that has been said, the cornerstone of good risk management practice can be seen as working to the standard of care. *Standard of care* is a vitally important concept. It doesn't mean you can blunder about blindly, but it does mean that you are not obligated to produce perfect work. Oversights, errors, or omissions may arise that could not have been anticipated or recognized by one meeting the reasonable standard of care in that place and time.

Earlier in this chapter, we stated that the doctrine of standard of care could be voided by faulty contracts or by offers to perform beyond the standard. You can foolishly eliminate standard of care as a defense if you guarantee or warrant your product to be perfect, or even to be of the highest quality or best level of service.

Some engineers will put marketing language above good risk management, telling prospective clients that they are "head and shoulders" above the other engineers in the area. Such bravado may come back to haunt you. It may be interpreted as a guarantee or warranty of a higher level of service than the "standard of care," and, worse, it may be an indefinable level. It may be whatever the client thinks it means, which is really bad for the engineer if a claim is based upon that premise. The court will assume that you knew what you were doing, even if the guarantee or warranty wording came in your client's form of contract.

I know of no professional liability carrier that doesn't exclude actions based on guarantees or warranties from the coverage. In today's competitive business climate, it is particularly difficult to object strenuously to an owner's form of contract at the risk of losing a commission, but both you *and the owner* have much to lose if your professional liability insurance is voided by the contract.

Risk management is a vast area of knowledge. Chapter 28 explores it in greater detail than we have here. Perhaps as a young engineer you would prefer to forget about the business elements of practice, but that would be unwise. There is much literature available on the topic of risk management for engineers, and many seminars are given by attorneys, insurance companies, and fellow engineers. Make it a point to gain knowledge in the area of risk management. It most certainly will include good design practices, good record keeping, and good management of personnel. But a well-run firm will also look to its managing engineers to be well versed in the matters discussed in this chapter. You will enhance your career dramatically if you start to learn about these principles almost as soon as you enter the profession.

Some prominent areas for risk management consideration can be better implemented if the areas of concern are anticipated at the outset of the project's contracting stage. Here is a partial checklist of commonly experienced problem areas.

Client Selection

- Does the client have appropriate and well-reasoned expectations for the project?
- Does the client have adequate financial strength?
- Does the client have a reputation for disputes and litigation with design professionals?
- Does the client have a good reputation in the industry?
- Does the client pay bills timely and in full?

Not all of this information will be readily available. There are, however, resources that can provide financial and other information on existing businesses. One is Dunn & Bradstreet (D&B), and another is the better business bureau. Both can be accessed through the Internet. Another resource comes through the networking opportunities you will have at your technical and professional society meetings. Perhaps you will meet others who have experience with the potential client and will be willing to share their general working experience with the organization or person.

CONTRACTS

One of the most basic forms of risk management is the preparation of good contracts. Although we discuss contracts in detail in Chapter 33, we spend some time here as well because contracts are critical to risk management. Oral contracts are permissible, but in terms of risk management, they are accidents waiting to happen. The best contracts are written, and at the top of the list are standard contracts generated by the Engineers Joint Contract Documents Committee (EJCDC). Engineers and/or clients can draft custom contracts, but custom contracts can introduce significant risk when either party attempts to unduly shift responsibility and liability to the other.

We will list some critical areas of concern in custom contracts, but we wish to emphasize that this book is not to be considered legal advice. We are reaching out to you as a young engineer in an effort to sensitize you to matters that may be of concern to you in the future and that we anticipate will enhance your career development. In a perfect world, all contracts will be subject to review by legal counsel and/or the engineering firm's professional liability insurance (PLI) carrier. Some common contract elements to consider are:

- Is the contract written?
- Is the contract signed?
- Is the signer for each side authorized to sign for the company? (Not everyone can sign legally, particularly in the case of corporations.)
- Does the contract require the engineering firm to perform to a level higher than that legally required (i.e., the standard of care to be expected of a similar engineer in the same location)? Expressions like "highest level of engineering service," "project will be fit for the intended use," or "plans will be prepared to obtain permit" are among the unacceptable risks. Those are contractual guarantees that go beyond the legal expectations of a professional engineer.
- Is the scope of service tightly defined?
- Is there a person with authority to act for each party designated?
- Are compensation and the payment schedule clearly defined?
- Is there a provision for renegotiation of contract terms if the client should change the scope of the project?
- Is it clear that the client will pay all application fees necessary for application for permits, review, and approvals sought on its behalf?

- Is there a provision for stopping work because of no or slow payment of invoices?
- Is there a first-stage dispute resolution forum specified prior to litigation (mediation preferred)?
- Is it clearly stated that the engineering opinions of construction cost (construction estimates) will be based upon readily available industry data and are not guaranteed to be accurate, because the actual cost will be determined by bidders and/or negotiations with contractors.
- Is document ownership to remain with the engineer (intellectual property)?
- Are there any broad indemnification clauses in favor of the client that would place undue liability on the engineer?
- Is it stated that any certifications required by lenders or others will not be within the scope of services and that any requests for certifications will be addressed at the time they are made? Certifications need to be made only for those things that are within the control and knowledge of the engineer. Some lender certification forms have language that goes well beyond the knowledge or control of the engineer.
- Is the contract assignable by either party to another entity?
- Which state laws will apply if a dispute arises?

Subconsultants

- Is the subconsultant's scope of services clearly defined in written contract form?
- Is the subconsultant's skill and reputation known to the engineer?
- Is the subconsultant adequately insured? Certificates of PLI and general liability insurance (GLI) should be furnished to the engineer, along with cancellation notice provisions.
- Is the subconsultant's physical plant and staffing adequate for the project?
- Is the subconsultant properly licensed?
- Are invoicing and payment provisions adequately defined? When does the engineer need to pay the subconsultant?
- Is the subconsultant acceptable to the client?

Understanding professional liability principles is an essential element of the practice of engineering. There is much to learn, and this chapter is intended to give you a basic foundation on the topic. The topic is worthy of continuous study throughout your career.

13

LEARNING
SOFT SKILLS

COMPETITION

You must continually monitor and contend with two kinds of competition. The first is *internal competition*—competition for promotions, better assignments, and so on. from peers within your organization. The second is *external competition*—your organization's competition for consulting work, research work, design work, or for market share.

We would like to think that by purchasing this book you are already one step ahead of your internal competition. One of the basic tactics from military training is to identify all your enemy's strengths and weaknesses, then exploit their weaknesses and not allow them the opportunity to use their strengths. This will enable you to defeat your enemies.

¡Cuidado! Be careful!

In the business world, you really don't want to vanquish the enemy; you just want to do a better job than the enemy. You need to be assertive, not overly aggressive or dangerously passive. Maintain a positive "can do" attitude. Evaluate your competition's weaknesses and turn out a recognizably better performance in these areas. Also, use all the suggestions about improving your visibility that were described in Chapter 2; this is really a form of marketing yourself.

You need to be assertive, which may be difficult for many young engineers. As a young, relatively inexperienced engineer, it may be difficult for you to get others to listen to your ideas and opinions. The key is to speak up and state your ideas or opinions, along with your reasons for reaching your conclusions. This is important in daily discussions as well as in formal meetings. If you are wrong, treat it as a learning experience. You will be respected for having opinions and ideas; you will be respected for listening and learning in the process.

To really understand your external competition, you may have to do some research. Fortunately, the Internet greatly facilitates this research. If you are in the design or consulting business, your competition is those firms that bid against you for jobs. If you are in industry, your competition is those companies that market similar products. It is easy to find their Web sites on the Internet to learn who they are. The Web site may include curricula vitae of the principals; this is a good indicator of their strengths. Word-of-mouth opinions can provide even more in-depth analysis, but these opinions can be biased.

Why should you do this research? You may learn ways to improve the operations of your organization. You will be able to represent your organization better when you meet potential clients. You will become knowledgeable about the special services of other firms that your organization has chosen not to perform. All this helps in your career development and improves your personal image.

MARKETING

A cornerstone of marketing is understanding your competition, but you can do a few other things as well.

Market yourself. Maintain positive visibility as previously discussed. You are a professional. Maintain a professional appearance. Polish your abilities with soft skills such as communications, writing, negotiating, being a good team player, dealing with stress, and developing basic management skills.

Extend this to promoting your firm, as well as carefully promoting yourself, to the world outside your organization. Remember that every time you meet with a client, a potential client, a government official, a regulatory organization, or a local business person, you are representing your organization. It is your obligation not only to professionally represent your organization but to do so in a positive manner.

Large organizations have marketing departments. Smaller engineering firms often have a marketing individual on staff. You should meet with these people periodically and learn about their programs. If they produce marketing materials, keep a small supply of these in your briefcase and use them judiciously.

Your business card is a marketing tool. It usually is formatted in compliance with the basic format of your organization. You should add special credentials such as an advanced degree, PE license, technical specialty, and the like. A number of Web sites provide free information on effective business card design concepts.

We often think of marketing as telling others all about the great products or services that we have to offer. But marketing involves two-way communications. First, we need to listen to others to learn both what they think is great and what they think could be improved about our products and services. Then we need to make adjustments and give feedback to these persons. Second, we need to scour our marketplaces constantly for ideas for new products or services or for a new niche. You will certainly impress your management if you bring back any such ideas that will improve your organization's profitability.

Marketing can be a career choice for engineers. Opportunities exist in large organizations for marketing engineers to provide design services used in conjunction with sales of equipment. The quality and cost-competitiveness of these auxiliary design services is a key ingredient of product sales. There are also opportunities for marketing engineers with firms that primarily sell lines of equipment where purchasing decisions are made by other engineers. Some examples are lighting equipment or HVAC equipment sold to building designers or plant managers.

It should also be noted that in small engineering firms, the principals of necessity wear two hats. They are the technical experts, but they are also primarily responsible for marketing the firm and its services. It is not easy to wear both of these hats at the same time. Nevertheless, the marketing efforts of the small firm are key to its survival and growth.

COMMUNICATIONS

A classic pun recommends that you should never ask an engineer what time it is, because an engineer will tell you what is inside the watch, how it was made, how accurate it is, and finally give you the time. This illustrates the fact that engineers often have problems with effective communications, both written and oral.

Many books are available on effective writing and effective speaking. For those organizations that have internal continuing education programs, these two courses are ubiquitous. Even if you think you are effective at both writing and speaking, you probably could use some improvement. This is especially true if English is not your native language. It is beyond the scope of this book to provide a treatise on effective writing and effective speaking, but we will highlight some points that are pertinent for engineers.

An engineer who can communicate clearly, concisely, and convincingly and also has a good technical understanding is usually recognized by management as an engineer with high potential. Engineers often write technical reports and specifications. Specifications often define requirements for products or processes. If the specifications are not clearly understood, delays, cost overruns, or even disasters can result.

For successful career progress, you need the ability to sell yourself and your ideas—to your supervisors and to your peers. You need to communicate effectively with other members of the design team. You also need to develop your ability to produce a good "elevator speech." When your vice president asks you on the elevator for a project status report, you need to deliver a concise but complete update in well under five minutes. If you do it well, without fumbling, you create a great impression.

In his now old, but still good, book *The Technique of Good Writing*, Robert Gunning (McGraw Hill, New York, 1952) introduced the fun concept of the "fog index." By measuring the average sentence length and number of words with three or more syllables and putting them into a formula, one can measure the "fog index" as an indicator of unintelligibility. It can also measure the amount of meretricious persiflage in a document. That sentence should have sent you to your dictionary! The point is that engineers, in their communications, tend to forget that their audience often is not other engineers. The success of forensic engineers (those that specialize in giving expert testimony in courts of law) is directly related to their ability to communicate their expertise is simple, layperson's terms.

In addition to courses and books, another way to develop communications skills is to watch other effective communicators in action. What makes them successful? What do they emphasize? How do they organize the presentation? How do they summarize and draw conclusions? Toastmasters International (*www.toastmasters.org*) and other such organizations offer great learning opportunities for communications skills.

NEGOTIATING SKILLS

Some people love to argue. Are they therefore good negotiators? Effective negotiations result in one or more parties changing their position based on convincing arguments during negotiations. Incessant argument, sometimes referred to as verbally beating down the opponent, does not necessarily result in successful negotiations.

We all negotiate on a regular basis without thinking about it. Where should we go to lunch? Should we buy a new lawn mower or a new dishwasher? Will we vacation this year at the shore or at the lake? We usually answer these questions through negotiation.

Many people shy away from negotiations because they view the process as adversarial and therefore uncomfortable. However, negotiations are a normal part of business activities. Engineers need to negotiate in project meetings about changes in project scope, when assigning project tasks, for due dates, when determining the best solution for engineering problems, for business contracts, and in many more activities.

It should also be noted that if your engineering assignments are multinational, the culture in other countries may make negotiations a requirement for many more issues. Generally, Asians and Hispanics bargain for what seems to us to be everything—negotiation is part of their way of life. If you are not sure of the negotiating practices of another culture, get help by consulting with an expert.

When negotiating, you identify your needs and then try to identify the other party's needs. This means doing your homework before the negotiating session. The American way usually is to split the difference, which hopefully produces a win-win resolution. For aggressive negotiators, and for some other cultures, the desired outcome is I win/you lose.

When negotiating, try to put yourself in the other party's shoes. Try to understand its needs. If you have to deprive the other party of some of its needs, structure a trade. You may also have to provide a maneuver or condition to help the other party "save face." The eventual desired outcome is for both parties to reach a satisfactory and robust settlement that covers all issues. Negotiating is an essential "soft skill," but it is also an art.

You can best learn negotiating in a formal course where there are practice sessions. You can also learn by sitting in and observing senior members of your organization conducting negotiating sessions. Learning to negotiate is a continuous process. Do a postmortem analysis of each of your negotiating sessions. Was your homework adequate? What were the positions taken by the other party? What were the characteristics of the other party? How could you have done things differently? Difficult negotiating sessions should be reviewed with your mentor or with an expert. Remember, negotiating is much like a game. Learn from each session, but have fun too!

TIME MANAGEMENT

H. Jackson Brown, Jr., noted author, said: "Don't say you don't have enough time. You have exactly the same number of hours per day that were given to Helen Keller, Pasteur, Michelangelo, Mother Teresa, Leonardo da Vinci, Thomas Jefferson, and Albert Einstein." So many of us have found ourselves thinking that we need more time to do all there is to do. Such thoughts drive some people to a sense of desperation, a feeling of being totally overwhelmed by what is on their plates. The answer: practice time management.

In 1987, an Italian economist, Alfredo Pareto, theorized that 80 percent of the land and income in his country was controlled by 20 percent of the populace. In later years, this principle was expanded by others to suggest that managers should spend 80 percent of their time on the most important 20 percent of tasks. The Pareto Principle is procedurally formalized, and you will find much written on its application and usage. It is important to know that a procedural method exists for identifying problem areas and setting formal procedures to analyze them for relative importance. Charts to graphically present the relative value of problems can be derived and then used to guide reducing, solving, or eliminating the most important problems. If a manager can identify the most important 20 percent of the tasks on the agenda, putting 80 percent of one's time and effort into that 20 percent makes good sense.

Quality control and time management have roots in the manufacturing sector, but the principles are applicable to engineering in all sectors. Producing a set of plans to build a bridge, an airplane, a car, a computer, a chemical plant, or a municipal road all demand quality control and time management to be effectively achieved on time and within budget. Time management is essential for all participants in the design project, especially for the project managers. For those engaged in the business management side of a design unit, time management is equally as important.

How can one focus on efficiency through time management? It is a daunting task. Everyone knows that deciding what to do during a day or week at the office is fairly simple in theory, but external events, the telephone, needs of staff or colleagues, meetings, clients or customers, unforeseen emergencies, and so on can and will disrupt the plan. Time management won't stop the external events but may help to navigate the shoals more effectively.

One of the vital parts to the time management process is the creation of a plan. To do that, you need to start with a list of known tasks. The first thing to do is to list everything that you can think of that will be necessary in the reasonably foreseeable future. For personal tasks, a week or two might be reasonable. Once the short-term technique has been mastered and becomes second nature, you should move on to more extended horizons. Perhaps the next stage would be to project personal and project goals for a three- or six-month period. For project planning, other members of the project team may be enlisted. For this exercise, the focus will be on personal tasks.

Once that list is prepared, including personal needs, such as "dentist appointment" or "take car for service," the real work begins. Tasks need to be prioritized. Remember the Pareto Principle. Try to find the 20 percent of tasks that are most vital. Each day, start out by looking at the short-term list. Look at where you are on the plan, what prioritized tasks need to be addressed, and plan the day. Keep priorities in mind, making an effort to avoid the minutia that comprise the 80 percent of less important tasks. Lower-priority work should be, to the extent possible, put aside or eliminated altogether. Perhaps some of the lower-priority tasks can be delegated to someone with less responsibility. For instance, seeking outside assistance from a vendor or a client/customer is something you really want to do, but it will take time to make the calls, be on hold, wait for a return call, and so on. In a lot less time, you can explain the needed input to a draftsman or to an administrative assistant and get back to your primary duties in engineering design or project management. The key to this is the plan. If the call was not in the 20 percent of highest priority, delegating the work should be an obvious choice.

A difficult part of time management is the need to control the unscheduled interruptions of colleagues or subordinates. At times, an interruption is worth handling. For instance, if a drafter is totally stymied in efforts to apply engineering sketches into a drawing, the delay in a project manager's day may be warranted, because the project will be on hold until the issue is resolved. If, however, a drafter wants to discuss an issue regarding the CADD technology, it is likely that it can wait until it fits into your scheduled time for such interfacing. Obviously, one should be diplomatic, but firm, in explaining the need to wait until a later time to discuss the issue.

A driving principle of time management is the reduction or elimination of wasted effort. It would be inhuman to suggest that anything not related to production is impermissible. A work environment devoid of interaction with fellow workers is not desirable or even achievable. After all, engineering is a profession requiring interaction between people—engineering is a team effort. Brief exchanges between co-workers are to be expected, even encouraged. But chatting at length about sports, movies, rumors about other firms in your business, or a myriad of other topics unrelated to the work at hand is not useful in time management and may even result in low ratings from supervisors. In fact, most engineering positions require one to fill out time records that are input to a project management accounting system, most of which have modules that allow review of the time spent on tasks by individuals. This is not to say that one's only motivation in mastering time management skills is to avoid a poor appraisal, but knowing that your supervisor may evaluate your time management should present another level of motivation.

A significant factor in time management is to avoid the tendency to perform a lot of small, simple tasks over a lengthy period of time while avoiding digging in to a major duty. Examples of small tasks could include dealing with e-mail; writing memos that can be deferred or assigned to others; making calls that can wait for hours, if not days; reading magazine or periodical articles that are not essential; working on short-term assignments that are fun and easy but not time sensitive; and more. Remember the 80/20 principle. Those tasks are probably not in the 20 percent that have been prioritized but they can easily consume larger chunks of your available time than you would like.

Effective time management is best achieved through a disciplined approach. You may need to identify and modify long-ingrained habits. A useful tool for improving time management is a calendar or diary. A variety of technology-based tools are available, such as Microsoft Outlook, PDAs, and other project-planning programs. Such technology-driven aids may be installed on your employer's network, and you are well served by learning to use those programs expediently. Not only should events be posted but some thought should be given after each event about the event's usefulness to one's primary objectives. For instance, there may be a weekly meeting with the project team along with upper management and other interested parties. Such meetings can often run long, with outcomes insufficient to warrant the time investment. One may not be able to change that, but a diary with time and outcome assessments may be sufficient to develop a compelling position to present to a supervisor. Perhaps the program will be adjusted because of the record. If not, the engineer who gave serious thought to the issue will have made a best effort, perhaps even getting positive attention from management. Of course, any such effort should be presented in a very respectful, professional manner and must be presented as an observation or recommendation made in the interest of bettering the organization.

As you advance in project management, the time management skills of your subordinates in a project team need to be assessed. If some are weak in these skills, they will hurt the team. If the project manager has become proficient in that area, then motivating others to manage their time effectively will be a lot easier, as will the ability to mentor them to do so. The skills and techniques of the mentor can be shown to the protégés with sincere conviction, demonstrating effective techniques. The results will be amazing.

An entire book could be written about time management—and many have. Young engineers should be aware of the wealth of information available to help develop this necessary skill. This section is meant to teach you about the need for time management and some of the pitfalls that interfere with the effective management of time. Remember the well-known adage, "Time is money." Most engineers are engaged in production of products or services. Their employers pay them, usually an hourly rate, to do the work of the organization. If you are distracted or inefficient in doing the work of the employer, that is a disservice at best, unethical and immoral at worst. Managing time is not only a commendable goal but also an inescapable duty.

GEMS OF CAREER ADVICE

You'll find a section titled "Clean Up Your Act" in Chapter 16 on outsourcing. It includes a short list of steps you can take to make sure your performance stays in the outstanding category, especially when your career may be threatened by outsourcing, downsizing, mergers, takeovers, or cost cutting because of economic downturns. Following are more tips, including some regarding soft skills, to help you maximize your performance for surviving these threats to ensure a successful career:

- Develop a formal career plan; link it to your personal vision.
- Know your strengths and weaknesses: learn what differentiates you from the crowd and market it, and fix your weaknesses.

- Show enthusiasm for your work.
- Maintain a high energy level in your work.
- Meet your deadlines; deliver on your promises.
- Develop and use a mentor.
- Find a champion (not the same as a mentor) who can help pull you upward in the organization.
- Always act and speak like a professional.
- Always dress like a professional and be properly groomed, following the dressing habits and mannerisms of your management.
- Seek diversity in your assignments; broaden your experience and become more valuable.
- Never stop learning; keep up with technology advancements.
- Never stop networking: your contacts are future resources, so stay in touch, be friendly, and be cooperative.
- Not all projects can be highly successful, so have backup plans.
- Try to select a supervisor from whom you can learn the most and who can pull you along on the fast track.
- Learn to deal with stress.
- Always maintain the highest possible degree of integrity.

14

ACCOUNTING AND CORPORATE FINANCE

I had a great plan for the new product I designed, but the accounting people said its internal rate of return was too low. What does that mean?

Adding this product, with it's proposed form of manufacturing, to the company's product line could have an adverse impact on "EPS." What does that mean?

I am considering a move to another company, but my brother-in-law, who owns a lot of stock, says that company's P/E ratio is too high. What does that mean?

Most engineers have not had courses in accounting or in engineering economics. (Exceptions are the industrial engineering, management engineering, or sometimes the construction management curricula.) In this chapter, we discuss some basic accounting concepts that will help you to understand the impact of your engineering work on corporate finance. This chapter is by no means a complete accounting course, but it will help you to understand what the accountants are saying. It may also help you better understand your status as a possible partner or shareholder in your organization. You may have the opportunity to become a shareholder through an employee stock option plan. A better understanding of corporate finance will help you evaluate such opportunities.

If you ever become involved in starting or expanding a small engineering practice, these concepts are also used in the required business plan. Many years ago, small engineering firms looked at their checkbook at the end of the year, and if there was a positive balance, it was assumed that all was copacetic. Now, income and other tax reports are required by government agencies, and some clients require a balance sheet as part of prequalification for bid selection. These clients may also require that overhead rates must be calculable and well documented. Furthermore, requests to bankers for a line of credit require demonstration of professional financial records and a demonstrated know-how for maintaining these records.

If you are an engineer in the academic world or in government, you may not experience as strong a need for understanding accounting, and some of the accounting practices in these environments are a bit different. But your engineering or research activities are certainly constrained by the budgeting process, which in turn is driven by the organization's financial planning and operations.

In this chapter we will review the following subjects:

- General accounting
- Key operating ratios
- Cost accounting
- Project control
- Value engineering

GENERAL ACCOUNTING

Regardless of your organization, activity, or size, all businesses report their activities in four fundamental financial statements:

1. The balance sheet
2. The income statement
3. The cash flow statement
4. The reconciliation of net worth statement

Each statement serves a specific accounting purpose, and there is an interlocking financial relationship between these four statements. It is important to understand the purpose of each statement and learn how to read each one. It is also important to comprehend the relationships between the statements and how changes in one statement can have an effect on the other three.

We will show simple examples of these statements and provide basic definitions to help you understand each statement's purpose.

If you use Quicken® or Microsoft Money®, which most people only use for managing their checkbook, you can produce cash flow and income/expense reports. If you track all your assets, investments, mortgages, loans, and credit cards, you can produce net worth reports. If you fully use Quickbooks®, you can produce the four basic reports and more.

The Balance Sheet

The balance sheet is a record of the company's financial structure. It shows the following:

- What the organization owns (assets)
- What it owes (liabilities)
- What is left over for the owners or shareholders after liabilities are deducted from assets (net worth, or equity)

The fundamental accounting equation for this relationship is

$$\text{Net Worth} = \text{Assets} - \text{Liabilities}$$

The balance sheet is a static picture—a snapshot—of the organization's financial structure as of the date on which it was compiled. The balance sheet for the day before or the day after could look radically different, depending on the financial transactions that took place on those days. Table 14.1 shows a sample balance sheet.

Note that the top half of the balance sheet is current assets and current liabilities; the adjective *current* implies that they are more liquid or more easily accessible. The bottom half is long-term assets and long-term liabilities.

TABLE 14.1 *Balance Sheet of a Small Corporation*

ASSETS

Current assets		
Cash	$ 53,000	
Accounts receivable	68,000	
Inventory	98,000	
Prepaid expenses	6,000	
Total current assets		225,000
Property, plant, and equipment		
Land, buildings, and equipment	192,000	
Less accumulated depreciation	−17,000	
Net land, buildings, and equipment		175,000
Other assets		2,000
Total assets		$402,000

LIABILITIES

Current liabilities		
Accounts payable, trade	$ 61,000	
Accounts payable, other	22,000	
Accrued expenses	8,000	
Short-term debt	15,000	
Income tax payable	7,000	
Total current liabilities		113,000
Long-term debt	75,000	
Total liabilities		188,000
Shareholders' equity		
Capital stock	150,000	
Retained earnings	64,000	
Total shareholders' equity		214,000
Total liabilities and shareholders equity		$402,000

The balance sheet—or a series of balance sheets covering a series of reporting periods (quarters or years)—is useful for forming an image of the company's basic financial structure and level of indebtedness at a particular time. It does not, however, directly reveal how the company is doing financially, whether it is making money in sufficient amounts in a particular period to meet its obligations (its liabilities) and increase the net worth for the company's shareholders.

If the organization went out of business, the value of the assets (which may require adjustment for actual sale amount) minus the debt (loans, accounts payable, etc.) is what the owners and/or shareholders would walk away with.

The Income Statement

The income statement is a record of the financial performance of the organization (its ability to make money or produce a profit) over a period of time. The income statement records all the revenue generated by the organization during the period and deducts all its expenses for the same period to arrive at *net income,* or the profit for the period. Net income can be a negative number—if total expenses exceed total income, the organization realizes a net loss.

Table 14.2 is an example of a simplified income statement.

The net income divided by the number of shares of stock outstanding is a key indicator called earnings per share (EPS). Large projects are sometimes evaluated to predict or measure their impact on EPS.

When the income statement shows a profit, you may think the profit is all available in cash or in a checking account to use for the operations of the business. Good assumption, but hold on a minute! If you carefully examine the income statement, you will see that not all income and expense items are cash. Certain items, such as depreciation, are recognized as expenses by generally accepted accounting principles, although they do not require an outlay of cash. Also, certain items may be recognized as income before cash actually flows into the company or as expenses before cash flows out of the organization.

An example is recognizing as income the shipment and sale of some of the company's products, although the buyer has not yet paid for these products. Thus, certain income and expenses are recorded when they are *accrued,*

TABLE 14.2 *Income Statement of a Small Corporation*

Sales	$1,020,000
Cost of goods sold	561,000
Gross income	$ 459,000
Operating expenses	278,000
Depreciation expense	17,000
Operating income	$ 164,000
Interest expense	11,000
Income tax expense	39,000
Net income	$114,000

not when cash actually flows. This is called *accrual accounting,* another accepted accounting principle. Accrued items also flow through from the income statement into balance sheet accounts. As a result, the income statement does not reflect the true cash position of the organization. That is the role of the cash flow statement.

Cash Flow Statement

Cash is the most liquid asset of a company; it is money in the bank and the green folding stuff of which we never seem to have enough. The cash flow statement reveals the amount of cash generated by a firm over a defined period. Cash outflows are subtracted from cash inflows to derive the net change in cash for the period. The cash flow statement tells us how much excess cash was generated by the business after meeting all cash expenses for the period. This net cash is the amount of money available for additional cash expenses, such as additional debt payments. Net cash can also go negative. If this happens, the organization is using cash reserves from prior periods to meet its cash expenses, and management is prone to kick off yet another cost-cutting program. If the negative cash flow trend isn't reversed, the organization will eventually run out of cash and you are likely to run out of a job! So pay attention!

For many years, cash flows were calculated informally by extracting information from the income statement and from certain balance sheet items. A variety of methods were used, some more accurate than others, and some organizations did not bother to produce cash flow statements at all. Ignoring cash flow history and projections led to problems during the 1980s, when businesses borrowed aggressively from willing lenders, then ran short of the cash required to make loan payments. This was known as the savings-and-loan scandal era. As a result, cash flow analysis standards and requirements were revised, and cash flow statements became a key financial reporting document.

The cash flow statement is similar to the income statement in that it records a company's performance over a specified period of time, usually over the quarter or year. The difference between the two is that the income statement also takes into account some noncash accounting items such as depreciation. The cash flow statement strips away all of this and tells you how much actual money the company has generated. Cash flow shows us how the company has performed in managing the inflow and outflow of cash and provides a sharper picture of the company's ability to pay bills and creditors and to finance growth. Many of the items on the cash flow statement are also found on the balance sheet or income statement, but here they're arranged to highlight how cash relates to reported earnings.

The cash flow statement is divided into three parts:

1. *Cash flow from operations.* Cash from operations generated from day-to-day business operations
2. *Cash flow from investing.* Cash used for investing in assets as well as the proceeds from selling parts of the business, equipment, or other long-term assets
3. *Cash flow from financing.* Cash paid or received from issuing securities and borrowing funds; may include dividends paid (although dividends are sometimes listed under cash from operations)

Table 14.3 is a sample cash flow statement. In addition to the categories described above, it also shows which items are a source of cash and which items are a use of cash. It also shows how the entries are related to changes on the balance sheet.

TABLE 14.3 *Cash Flow Statement*

Cash flow from OPERATIONS		
Cash receipts from sales (Sales − accounts receivable)	$952,000	Source of cash
Cash payment for product (CGS + inventory − payables, trade)	(606,000)	Use of cash
Cash payment for operating expense (Operating expense + prepaid expense − accounts payable, other − accrued expense, operating)	(260,000)	Use of cash
Cash payment for interest (interest expense − interest payable)	(9,000)	Use of cash
Cash payment for income tax (Income tax expense − income tax payable)	(35,000)	Use of cash
Total cash flow from OPERATIONS	42,000	
Cash flow from INVESTING		
Property, plant, and equipment	(192,000)	Use of cash, asset increase
Other assets	(2,000)	Use of cash, asset increase
Total cash flow from INVESTING	(194,000)	
Cash flow from FINANCING		
Short-term debt	30,000	Source of cash, liability increase
Long-term debt	75,000	Source of cash, liability increase
Capital stock	150,000	Source of cash, liability increase
Dividend payments	(50,000)	Use of cash, net worth decrease
Total cash flow from FINANCING	205,000	
Net change in cash	$53,000	

The Reconciliation of Net Worth Statement

Reconciliation is what you do after you insult your spouse, right? What does it mean in accounting? Generally, accountants use a formal process to make sure the numbers in one kind of report agree with the numbers in another kind of report. You do it when you go through the exercise of making sure your checkbook balance agrees with what the bank says that you have. The reconciliation of net worth statement informs us of the changes in the net worth of a business during a financial reporting period. It shows:

- How much net worth increased or decreased as a result of net income or net loss
- What distributions (dividends) were made to shareholders (the business owners)
- What additional funds were invested in the business by shareholders, perhaps through the sale of more stock or by owners transferring personal funds to the business

Here is a simplified reconciliation of the net worth statement:

Beginning balance	$250,000
Plus net income	125,000
Less dividends	75,000
Ending balance	$300,000

In the case of a large company, this statement is called the Changes in Stockholder's Equity Statement. Today we hear the phrase *growing the company*. Stakeholders (shareholders, owners, employees) are not happy unless the company is growing. Growth is measured by the ending balance being consistently bigger than the beginning balance on the reconciliation of the net worth statement.

KEY OPERATING RATIOS

Ratios of numbers from various portions of the accounting statements communicate various information. For example, operating ratios are used as a measure of performance of the company's operations, and investment ratios are used by investors to try to predict future performance of the company. We have selected five ratios for further description. These are not necessarily the best five out of all the ratios, but they seem to be popular.

You certainly do not need an intimate knowledge of these ratios for your engineering work, but they are helpful to evaluate the economic status of your current employer as well as possible future employers.

Ratio analysis isn't just comparing different numbers from the balance sheet, income statement, and cash flow statement. It's comparing the number against the previous years' performance, the performance of other companies, the average performance of the industry, or even the economy's performance in general. Ratios look at the relationships between individual values and relate them to how a company has performed in the past—and how it might perform in the future.

How do you know if a given ratio is good or bad? It's usually more useful to compare the ratios of one company to other companies in the same industry, to the market in general, or against the company's own historical trend for the given ratio. Several Web sites and numerous investor services enable you to review industry norms and ratios to see if your company's numbers are above, below, or equal to others in the same industry.

Current Ratio

Current Ratio = Total Current Assets ÷ Total Current Liabilities

The current ratio is regarded as a test of liquidity for a company. It expresses the "working capital" relationship of current assets available to meet the company's current obligations. Current assets and current liabilities are taken from the balance sheet.

Return on Sales, or Profit Margin (%):

Profit Margin = (Net Profit ÷ Net Sales) × 100

The net profit and net sales are taken from the balance sheet and income statement, respectively. The ratio measures the percentage of profits earned per dollar of sales and thus is a measure of efficiency of the company. The profit margin of a company determines its ability to withstand competition and adverse conditions like rising costs, falling prices, or declining sales in the future.

Return on Equity, or Net Worth

Return on Equity =
(Net Profit ÷ Net Worth or Owners Equity) × 100

Net profit and net worth are taken from the income statement and balance sheet, respectively. A company's return on equity measures the ability of its management to generate adequate profit for the capital invested by the owners. Generally a return of 10 percent is desirable, both to provide dividends to owners and to have sufficient funds for future growth of the company.

Return on Net Worth

Return on Net Worth =
(Net Profit after Taxes ÷ Net Worth) × 100

The net profit after taxes and net worth are taken from the balance sheet and income statement. This ratio is sometimes called return on equity. It analyzes the ability of the firm's management to realize an adequate return on the capital invested by the owners of the firm. A variation of this is the internal rate of return, which evaluates the same quantities but only for a specific project.

P/E Ratio

The P/E ratio is widely quoted in stock analysis. It is the price per share of the stock divided by the earnings per share.

In general, a high P/E suggests that investors are expecting higher earnings growth in the future compared to companies with a lower P/E. However, the P/E ratio doesn't tell us the whole story by itself.

The P/E is sometimes referred to as the "multiple," because it shows how much investors are willing to pay per dollar of earnings. If a company is currently trading at a multiple (P/E) of 20, the interpretation is that an investor is willing to pay $20 for every $1 of current earnings per share.

COST ACCOUNTING

Cost accounting is used in all types of organizations to determine the actual costs of products and services. Cost accounting is sometimes limited to determining production cost. Because many organizations produce more than one product or service, they need to allocate certain overhead costs across several different products. In the engineering firm or in university research, the "products" are the provision of services in the form of projects.

Costs directly assigned to projects include labor, other direct costs incurred such as travel and reproduction, and payments for consultants or subcontractors. Time sheet accounting permits allocation of labor costs to projects, and other direct costs generally are assigned through a system of coding these costs to projects when they are paid. The engineering firm must allocate to projects, on some equitable basis, all other costs that cannot be assigned directly. These are the overhead costs, including office space costs, computer costs, taxes, insurance, and so on. Because direct labor (time charged to projects) generally is the most significant cost item, it often is used as the basis for allocation of overhead costs.

Ideally, all labor costs should be charged to projects, resulting in all labor hours being billable hours. In the real world, some labor hours are not attributable to specific projects and therefore effectively become overhead costs.

Cost accounting is frequently used to facilitate internal decision making and provides tools with which management can appraise performance and control costs of doing business. Product or project cost accounting is often an independent analysis that uses many of the same source documents as corporate accounting. Product or project cost accounting is used to control projects and is not directly used to provide information to the organization or corporate financial statements.

PROJECT CONTROL

There are three major categories of project control:

1. Project schedule control
2. Project cost control
3. Project performance control

Modern project control software integrates all three categories into one program. In this chapter, we review only project cost control because it is so closely related to cost accounting.

Project control uses the cost accounting methodology to compare project costs to project budgets. Whether a project is being administered on a lump-sum (fixed-fee) or on a time-and-materials basis, firm management needs to track project costs to determine overall profitability.

The costs incurred on a project are from labor hours, equipment, materials, contract costs, and overhead. The first three are collected or measured by field personnel. Labor cost is procured from time sheets and reported by activities that were defined in the project estimate. Equipment costs are procured from invoices and field supervisor's reports on hours of operation. Material costs are procured from invoices or delivery tickets. All of these costs are apportioned to the proper activities and compared with the activity budget.

A complete analysis of project control and project management is beyond the scope of this chapter. However, many books and courses are available on this subject.

VALUE ENGINEERING

Value engineering (VE) is an organized approach to providing the necessary functions of a product or project design at the lowest cost. It is also an organized approach to the identification and elimination of unnecessary cost. Unnecessary costs are those that provide neither use, nor life, nor quality, nor appearance, nor customer features to the product or project. Administered properly, value engineering does not affect the quality of the product. VE deals with component costs and, therefore, is often confused with cost accounting.

Value engineering is the systematic application of recognized techniques, often by a multidisciplinary team, to identify the function of a product or service, establish a worth for that function, generate alternatives through the use of creative thinking, and provide the needed functions to accomplish the original purpose of the project at the lowest life-cycle cost without sacrificing safety, necessary quality, and/or environmental attributes of the project.

VE studies do all of the following:

- Use an independent technically diverse team
- Follow a systematic job plan
- Identify and evaluate function, cost, and worth
- Develop new and unusual alternatives for required functions
- Determine the best and lowest life-cycle cost alternatives
- Develop fully supported recommendations

Projects that have already experienced cost, schedule, or scope problems benefit from VE analysis. But the greatest potential for improvement is in technically and organizationally complex or unusually constrained projects in preliminary design (20–35 percent completion). VE at this point produces maximum benefit because recommendations can be implemented without delaying progress or causing significant rework of completed designs. While the average cost improvement from VE is 6 percent, cost reduction is not always the most significant benefit. Schedule reductions, environmental requirement modification, and operational procedures can all be improved through the functional cost evaluation used in all VE studies.

SUMMARY

This chapter provides information to help you better understand business operations. This material becomes more meaningful as you move up into a supervisory or management capacity. Copious information about corporate finance is available in textbooks and on the Internet. You may be interested in investigating some sections of this material in greater depth. Should you decide to obtain an MBA, you will learn much more about corporate finance. Just remember, you were hired for your engineering knowledge and experience; you can leave the daily accounting details to the bookkeepers and the accountants. But remember also that you need to be conscientious about any finances for which you are responsible. Your careful budgeting and spending has an impact on the company's finances.

15

MOONLIGHTING

Many young engineers seek out opportunities for work outside of their regular employment, either to gain new experiences at a faster pace or to supplement their income. This common practice is often called "moonlighting." For engineers, some special considerations are associated with the practice of moonlighting.

Some moonlighting jobs may be unrelated to an engineer's regular work. For instance, if you were to work part-time at a mall as a salesperson or as a member of a musical group, or do any other sort of work not involving engineering, there might not be any issue of concern. However, if the moonlighting work caused you to be fatigued at your regular job, fail to meet regular work hours, or be distracted during the workday, there is an ethical problem. If you were to moonlight within the engineering industry, there can be even bigger issues. They include:

- Ethics as an employee
- Duty to employer
- Duty to client
- Personal liability exposure
- Liability exposure to your employer

ETHICS AS AN EMPLOYEE

In Chapter 11, we explored the meaning of ethics and pointed to the codes of ethics of several organizations. We pointed out the close relationship of those codes of ethics with laws governing the practice of engineering. We truly hope that we have adequately prepared you to practice ethically and legally. In the appendix, we have furnished copies of the codes of ethics of the National Council of Examiners for Engineers and Surveyors (NCEES) and of the National Society of Professional Engineers (NSPE). It should be noted that while we will refer to the NSPE Code of Ethics here, many professional and technical engineering societies promulgate codes of ethics. You might wish to look at other codes such as those of the American Society of Civil Engineers (ASCE), the American Society of Mechanical Engineers (ASME), the Institute of Electrical and Electronics Engineers (IEEE), the American Institute of Chemical Engineers (AIChE), or the American Institute of Mining, Metallurgical, and Petroleum Engineers (AIME). You can also visit the Web site of the Online Ethics Center for Engineering and Science (*http://onlineethics.org*) for other links.

Let's look at a couple of specific provisions in the NSPE Code of Ethics that are particularly relevant to the practice of moonlighting.

III. Professional Obligations

1.c) Engineers shall not accept outside employment to the detriment of their regular work or interest. Before accepting any outside engineering employment, they will notify their employers.

6.b) Engineers in salaried positions shall accept part-time engineering work only to the extent consistent with policies of the employer and in accordance with ethical considerations.

Clearly, it is deemed unethical to moonlight to the detriment of your regular employer, and the code requires you to notify the employer of any such activity. It is also clear that you should not moonlight unless it is permitted by your employer's policy.

The code articulates these constraints for many reasons. Some of them will become evident in the following discussion. Bear in mind, however, that moonlighting has been carefully considered and addressed by the codes of ethics of the NCEES and NSPE, as well as other engineering societies.

DUTY TO THE EMPLOYER

In an employment relationship, the employer and the employee have duties and obligations to each other. Among your duties as an employee are loyalty, regular attendance, conformance to the employer's policies, practice of risk management (Chapter 28 deals with this topic in detail), avoidance of conflict of interest situations, and application of skills and knowledge gained on the job to the benefit of your employer. If you ever consider moonlighting, you need to do so in light of these duties.

Loyalty to Your Employer

Loyalty to your employer is a broad duty. It means that you will not perform acts that will be to the detriment of your employer. You should not accept moonlighting assignments that will lead to the use of your employer's proprietary information, or that will involve the application of special skills and knowledge acquired through your employer for use within the firm. Using your regular employer's equipment for the part-time employer would be disloyal, even after hours. Performing any of your moonlighting tasks while on your regular employer's premises, or even off the premises if you were on the job, would be disloyal. Accepting a moonlighting engagement without telling your employer would be disloyal.

Working for your regular employer's competitor would be disloyal. Working on your own time for one of your regular employer's clients would be disloyal, especially if you are doing it to help the client get by at a lower rate than would be charged by the employer for your time. In fact, accepting such a contract would be exceptionally dishonest, maybe even criminal. In my career, I have been invited to do that by my employer's clients, and I refused.

Regular Attendance

Regular attendance means that you will show up on time, be on the job during the designated work hours, and miss work only for acceptable causes, such as illness, family or religious events, permitted vacation and personal days. If moonlighting causes you to be late for work, to leave early, or to work ineffectively because you are fatigued or preoccupied by the moonlighting work, you are failing to meet a primary obligation to your employer. If a crunch in the moonlighting engagement causes you to miss work from your regular employer, you are not meeting your duty to your employer. If, while at your regular place of employment, you accept or make phone calls, send or receive e-mail messages, write or receive and read letters, all in relation to your moonlighting engagement, you are not in regular attendance. Regular attendance not only means being physically at your regular place of employment; it means being there mentally as well.

Conformance to Your Employer's Policies

Conformance to your employer's policies starts with your knowing what they are. Most firms have an employee policy manual. Some firms even ask every employee to sign the policy manual as certification that he or she has read and understood it. Moonlighting should be covered in the manual. If you cannot locate a manual, you should ask if the firm has any policies you should understand and, in particular, if a moonlighting policy is in place. The majority of design professional firms forbid moonlighting by employees. Others will discourage the practice but not prohibit it. Among the latter, many will require notification disclosing the potential client and the nature of the work. The employer reserves the right to say no if they have reason to be concerned. Many will require you to submit evidence of a professional liability insurance (PLI) policy covering your professional work.

Practicing Risk Management

To practice risk management in protection of your regular employer, you need to be able to identify potential hazards and, to the extent humanly possible, you need to take steps to minimize or eliminate those risks. If you accept a moonlighting assignment, it is not inconceivable that sometime down the road, someone may file a claim for damages allegedly arising out of your work. If that happens, the claim can sometimes spill over to your regular employer, especially if you have no PLI. That is why it is imperative that you identify and comport with your regular employer's policy on moonlighting, and that is why, if you do moonlight under that policy, you should carry PLI for yourself.

Avoidance of Conflict of Interest

Avoidance of conflict of interest was covered pretty well previously. To help emphasize the point, here are a few key conflict scenarios that you should avoid:

- If you moonlight for your regular employer's competitor, you are giving that firm the benefit of training you have received during your primary employment.
- If you show a competitor better ways to do things because you learned them at your regular employer's shop, you are in conflict of interest.
- While moonlighting for a competitor, you may inadvertently and innocently share information that will give negative ammunition to the competitor. The competitor may decide to spread the word that your regular boss is underpaying you and that's why you have to moonlight. It may not be so, but some people fight dirty in business.
- If you work for a client of your employer on your own time, you are perilously close to picking your regular employer's pocket.

Application of Skills and Knowledge

Application of skills and knowledge gained on the job to the benefit of your employer is pretty simple. As a former employer, I can tell you that my firm spent a lot of time and money in training new employees in many skill areas. One, in particular, was enormously costly. We were using a particularly versatile CADD program that required intensive training. We were among the firms that started using it soon after its release. We sent people to special training classes and conducted intensive in-house training as well. We also kept purchasing-enhanced releases of the software, only to repeat the training process for each release. It would have been a bitter experience to have our people moonlight for the competition and assist them in learning the program.

We also had some sophisticated surveying equipment, including Global Positioning Satellite (GPS) equipment. The same situation ensued. We sent people out of state for a week's training at the vendor's training facility. It would not have been well received if those folks used that expensively attained knowledge to help out our competitors.

Let me take that one step further. We are all familiar with athletes who get drafted into professional sports, spend the years of their initial contracts becoming superior players because of superior coaching, and then become free agents and leave the team, sometimes going to a competitor in the same division. We, as fans, may be discouraged when that happens, but we also recognize that the player has the right to do that. But what if a player, while still with his original team, had a part-time consulting job with a competitor to coach its defense against his team's offense? You can't dispute the wrongness of that. He would be using the skills and knowledge gained while working for his regular team to help the competition.

DUTY TO CLIENT

If you find that your regular employer's policy does allow moonlighting, you will also have a duty to your outside client to furnish quality service and to stand behind it. That means more than merely acknowledging errors and omissions, should any occur. It means having the fiscal resources to make the client or any injured third parties whole, if they can show that their injury or damages arose from your work. That is why you need to seriously consider PLI, and that is why your regular employer is doing you a favor if it requires it in its employment policy manual.

You will have duties to your outside client similar to those discussed above for your regular employer. You can appreciate that a definite potential for conflict exists if your duties to your two (or more) masters collide. For instance, if you have to attend a municipal board meeting to present your moonlighting work for consideration, and your regular employer needs you at a meeting at the same time, you will need to find King Solomon to help resolve it. Because of the possibility of such incidents, you would be well served if you let your outside client know at the outset that your primary duty is to your regular employer and that if there is a tie, your regular employer wins. Needless to say, if you read Chapter 33 and the section in Chapter 12 on contracts, you are very aware that you should reduce such understandings to writing.

PERSONAL LIABILITY EXPOSURE

A moonlighting engineer will be exposed to personal professional liability. When you work as a full-time employee, you should be covered by your firm's policy. If you moonlight for another engineering firm as a part-time employee, you may be covered under its professional liability policy. If you work as a contract employee, however, you may not be covered. You need to know.

If you are working for a private, nonengineering client using your own seal and signature on plans, your profile will be significantly raised. You will be much more prominent than you would be as an employee at your regular place of employment. While it is rare for employees of engineering firms to be named individually in claims, the signer and sealer of documents is more vulnerable than a "board engineer" who contributes to the project but is not in "responsible charge." If you moonlight for a private, nonengineering client, presumably you will be in responsible charge. By the way, if you are not licensed and you provide engineering services to an individual, it is likely that you will be breaking the law.

If you are going to sign and seal engineering documents for an individual client, I strongly recommend that you obtain PLI. It is a must. Even if you were to form a corporation to shelter you from other liabilities, professional negligence claims are personal in nature, and the corporate veil will not protect you. You need to assess the risk/reward ratio carefully if you intend to moonlight for individuals.

LIABILITY TO EMPLOYER

The primary reason that most employers will not allow moonlighting by employees is that they can be at risk from claims against a moonlighting employee. That is why your employer's risk management scheme may include an employment policy that will require you to disclose your potential client and the nature of the work and may require you to furnish proof of PLI insurance covering you before approving the engagement. That is why the employer's employment policy will tell you that you can't do any of that work on its premises, on its time, or using any of its equipment, telephones, e-mail, mailing facilities, copy machine, and so on. Your regular employer's

PLI covers you for claims arising out of work you do for the firm as an employee but not for work you do for others on a moonlighting basis.

I had a very bad experience at the firm I owned and operated for almost 30 years. We had a moonlighting policy that discouraged the practice but allowed it under certain conditions. One of my surveyors completely ignored the policy and did regular moonlighting on weekends. He worked for a local architect, a one-man shop. In fact, to make matters worse, he was using our equipment in the field and was using office equipment when no one was around to observe his actions.

I knew nothing about the illicit and unauthorized activity until I received a telephone call from an architect who had replaced the original architect on a project for a wealthy property owner in an estate community. It seemed that the topographic survey he had inherited from the original architect, the one-man shop fellow, was very inaccurate and was causing problems with the project. He learned from the first architect who the surveyor was and that he worked for my firm during the normal workweek. I explained that the work was done outside of our offices and was not supervised or billed out by us.

His client, the property owner, called me and demanded that I furnish him with a proper survey for free. I refused, of course. He threatened to sue my firm for failure to properly supervise my staff. He might have won, but he never did file the suit. I know firsthand what moonlighting can precipitate, especially if the moonlighter has no personal PLI. Needless to say, our unauthorized moonlighting employee was discharged immediately with cause.

We hope that your engineering career will furnish you sufficient professional and financial reward to preclude your needing to moonlight. But if not, or if you want to do it for the added experience, please follow our guidelines closely as well as the guidelines outlined by your employer.

16

OUTSOURCING

Discussions with colleagues around the water cooler often turn to the "foolhardy, inconsiderate practice" of outsourcing engineering work. It seems that most engineers know at least a few colleagues who have lost their jobs to outsourcing.

Outsourcing is one of the forms of competition you must deal with in your engineering career. To deal effectively with any form of competition, you need to learn as much as you can about it.

Outsourcing can take one or more of the following forms:

- *Consultants.* Projects or portions of the engineering operations can be assigned to consulting firms. This results in current employees being reassigned or terminated. The primary advantage to the company is a decrease in payroll and overhead costs. All engineering may be eliminated because it no longer can be justified as an ongoing profit center.
- *Offshoring.* Projects or portions of the engineering operations can be reassigned by brokering work to foreign service providers. This results in employees being reassigned or terminated. The primary advantage to the company is lower wages and elimination of the cost of fringe benefits. Sometimes tax incentives are a factor.

- *Temporary employees.* Projects or portions of the engineering operations can be assigned to "job shop employees." The primary advantage to the company is a reduction in fringe benefits' cost as well as the ability to hire and fire at will.

The use of consultants and temporary employees has been a practice for many years. It is a management technique to cut costs but may also result from reorganization. It is more understood and accepted than is offshoring.

THE CAUSES OF OFFSHORING

Whereas outsourcing is a form of competition to you, it is a result of another form of competition—the fiercely competitive environment in our global economy.

As an example, consider for a moment the garment industry in the United States. In the early 1900s, it was centered in New England. Then it moved to the South for lower wages, and the New England factories shut down. Next, the industry moved to Mexico for lower wages. Then it moved to a few Asian rim nations for still lower wages. Today, if you look at the labels in clothing you purchase, you'll notice that garments are made almost anywhere in the world, except for the United States. In the middle of this process, the U.S. garment market developed a high demand for designer garments that seemed to be indifferent to unit price. But the international entrepreneurs ignored U.S. trademark laws and flooded the market with high-quality reproductions. One can go to most any third-world country and purchase designer garments on the open market for a few dollars.

The globalization of work tends to start from the bottom levels of the workforce and move up the value chain into white collar and even management positions. The first jobs to be moved abroad are typically simple assembly tasks, followed by manufacturing and, later, skilled work like computer programming. At the end of this progression is the design work done by engineers and, finally, research and development work done by scientists and engineers.

The fiercely competitive global economy enabled the entry of millions of well-educated players from China, India, and the former Soviet Union onto the new playing field. The economic landscape has been completely altered with the proliferation of the outsourcing business model. How does this affect you? Engineering services in many sectors have been subjected to these same forces!

Offshoring is a relatively new buzz word, but the concept has been around for many years. The large consulting engineering firms have been operating internationally for many years. They have established operations in multiple foreign countries; these operations often employ a mix of local engineers and expatriates (U.S. engineers on temporary assignment). The same is true, to a lesser extent, for large industries.

An important reason for outsourcing engineering work is cost savings. The United States has the highest fringe benefits cost, a key component of overall labor costs. Perhaps the biggest portion of fringe benefits cost is the expense of providing U.S. employees with health care, which poses a very complex problem. To control total labor costs, U.S. companies are exploring or enacting outsourcing strategies. And the remaining U.S. workers are being asked to fund an ever-increasing portion of their fringe benefits costs.

As an example of cost savings from the outsourcing of engineering work, it has been reported that a top-notch engineering graduate with an MBA would start out in India at between $14,000 and $22,000 a year, while a U.S. engineering student graduating from a top management school would earn about $90,000 to $100,000 a year. The wage cost in India is just about a fifth or sixth of U.S. levels.

The outsourcing of engineering work is only a small part of business process outsourcing (BPO) activity. BPO has gone from a specialist capability employed by a few leading business organizations to a mainstream model for many companies. The Gartner Research organization estimates that the worldwide BPO market was worth $110 billion in 2002 and will grow to $173 billion in 2007, at a 9.5 percent compounded annual growth rate (CAGR). Offshore BPO services—where work is sent to another country—will grow from $1.3 billion in 2002 to $24.3 billion in 2007 at an 80 percent CAGR. Offshore BPO will grow to represent 14 percent of the total BPO market in 2007.

Outsourcing is a complicated process, so most firms retain the services of specialists called BPO service providers. Many of these consulting firms advertise on the Internet.

THE CONSEQUENCES OF OFFSHORING

The real impact of offshoring is debatable. Opponents see thousands of U.S.-based positions going overseas to individuals whose hourly rate is much lower than that of American workers. Opponents question the ability, and perhaps the willingness, of firms to assure the quality of engineering work performed overseas. Also, despite contractual assurances, opponents question the real ability to control the theft of intellectual property.

Opponents also question the legality of engineering work produced offshore that is not performed under the direct supervision of a professional engineer, as required by state regulations. Structural engineers in Georgia convinced the state licensing board in 2005 to clarify rules on the oversight of offshored work, says Jim Case, president of Uzun & Case, Atlanta, and former president of the Structural Engineers Association of Georgia. "We wanted to make sure you couldn't outsource and still be in compliance just by stamping drawings," Case said, regarding the proper responsible charge of outsourced design work.[1]

The advocates of offshoring contend that it is necessary because of the inadequate supply of engineers in the United States. Numerous studies show a decline in engineering graduates in recent years in the United States and a concurrent increase in graduate engineers in several other countries such as China and India. The supply and demand situation for engineers has been cyclical for many years. The advocates claim that the engineers used by offshoring companies are "highly degreed" and well supervised. In addition, advocates claim that offshoring creates high-paying management jobs in the home country and develops a new crop of entrepreneurs.[2]

Assuming that your management is being a Benedict Arnold when it announces outsourcing is a knee-jerk reaction. You must realize that there are no villains in this offshoring business model. Employers are reacting rationally to the evolving global marketplace. Employees have also been acting rationally by seeking alternate employment if they become victims. Per-

---

[1.] ("As Cost Pressures Mount, Offshoring Is Making the Work Go Round," Debra K. Rubin, Peter Reina, Mary B. Powers, and Tony Illia, *Engineering News Record,* 2 August 2004).

[2.] *Ibid.*

haps this lesson has not been forgotten from the closing of the buggy whip factories many years ago.

WHAT YOU CAN DO ABOUT OUTSOURCING

The first step is to stay current with the concepts of outsourcing as they evolve. Keep researching and asking about them until you understand the rationale behind them. Be sure to keep current on how the outsourcing concepts are affecting your industry and your company. To do this, you will have to monitor the business sections of your newspapers. You can also subscribe to a news service on the Internet and add "engineering outsourcing" to your search list. On the Internet, *www.forrester.com* has in-depth research reports, but they are not free.

It is difficult to dig out the information about your organization's plans regarding outsourcing. Your top management is reluctant to divulge information about studies and plans on this subject for obvious reasons. You need to pay attention to the rumor mill and use your internal resources to try to verify rumors.

If you think your job may be at risk because of outsourcing (or because of a merger, acquisition, poor financial performance, drop in revenues, loss of contracts, plant shutdowns, etc.) you need to get busy. Start with your supervisor. Make it clear that you are concerned about your future, especially because of things you have heard or read. If you get a quick, "You have nothing to worry about," probe a bit deeper. Ask about the security of your department. Is there sufficient work? Has outsourcing been considered? Have layoffs been considered? What is the future of your department? Should you begin looking for transfers to other departments that seem to have a better future? Which departments?

Because your supervisor may not be at liberty to reveal much information, or because your supervisor may not really know any details about future plans of your organization, you need to keep probing. Check with other contacts. Consult with your mentor. Talk to the person who does budgets for your department—what do the budgets assume in the future? Visit with your program manager and ask the same kinds of questions.

In doing all this research, remember to be polite and politically tactful. Don't be a tenacious nuisance to others. In other words, don't be "like a dog on a bone."

After you collect all this information, it is time to make a careful analysis. List all the positives, list all the negatives, and assign a reliability factor to each of your findings. Use this data to make the decision to stay and ride out the storm—or to start looking for a new job.

CLEAN UP YOUR ACT

If you decide to stay at your current position despite rumors, you may need to take steps to improve your performance. You need to ask yourself some tough questions. Is your performance rated above average? Can it be improved? Is your work of qood quality with little or no errors? Is your work getting good visibility with upper levels of management? When was the last time you volunteered for extra assignments? Are you working extra hours without being asked, and are you sure your management is aware of this? Are you willing to help other engineers in their work but not at the expense of the quality and quantity of your own work? Are you perceived as a team player? If not, what can you do to change that perception?

If there is a required workforce reduction (downsizing), remember that your supervisor may be the one who decides whom to keep and whom to let go. You need to make that decision to be easy, and in your favor, by being visible and turning out a consistently good performance.

You must also evaluate the type of work that you do. You may have become what is called a *commodity engineer*. With the increased power of computers today and the increased speed and reliability of telecommunications, any work that involves only formulas or design rules (sometimes called cookie-cutter engineering) can be easily outsourced at a lower cost and even performed 12 time zones away. If we look at left-brained (routine and detail oriented) and right-brained (context or big picture oriented) theory, commodity engineers tend to be left-brained. To avoid being a commodity engineer, you need to steer your career toward right-brained engineering work. These are activities that are nonroutine and require broad thinking, innovation and synthesis, customer/client development, project management, team leadership, and product management. Regarding innovation, you may not be an inventor, but you should regularly find new ways to do your job better, faster, and cheaper.

Therefore, you need to polish your distinctiveness. And keep in mind that in the global economy, your competition can be global in scope. How does one distinguish oneself when competing with a culture (such as China's) where if you are one in a million, mathematically that means there are still 1,300 people just like you?

BE AN ANTI-OUTSOURCING CRUSADER?

If the outsourcing business model has become an important issue to you, you may wish to do something about it. Especially as it applies to engineering, outsourcing by offshoring needs more study to answer questions such as:

- What is the impact on protection of intellectual property?
- What is the impact on national security?
- How will it affect free trade agreements?
- How can better information be procured about the number of jobs and the dollar value of the work that is outsourced?
- What is the relationship of outsourcing to the H1-B and L1 visa program?
- How can the problem with "responsible engineering charge" be resolved?

To contribute to the resolution of these problems, we suggest that you contact your engineering professional association. Many associations already have groups working on these problems.

Part Three

PROFESSIONAL DEVELOPMENT AND ADVANCEMENT

17

THE LICENSURE PROCESS

You have read about licensure as a professional engineer in multiple sections of this book. We are licensed professional engineers and have been for decades. We have encouraged you to become licensed for many reasons. In this chapter, we describe the process and the basic minimum requirements for you to attain licensure. It is a time-consuming process, and it will cost you some money. In the long run, however, the career enhancements derived from being a licensed professional engineer will far outweigh the costs of time and money.

The generalized steps given here are essentially true throughout the nation and its territories, but there will be state-specific variations. You will need to learn about those when you begin the process in any particular state.

A Web site you will want to visit is that of the National Council of Examiners for Engineers and Surveyors (NCEES) at *www.ncees.org*. There you will find much useful information about the licensure process, as well as many other topics, in addition to links to the various state registration board Web sites. We cannot overstate the fact that rules and regulations can change over time, so check this Web site periodically while you're gathering information on licensure as well as when you're actually ready to pursue your license.

The pathway to licensure can be defined by four mileposts:

1. Graduation from an accredited engineering program
2. Taking and passing the FE (fundamentals of engineering) exam, also known as the EIT (engineer in training) exam
3. Attaining four years of engineering experience satisfactory to the board of registration to which you will apply
4. Taking and passing the Principals and Practice of Engineering exam (PE exam)

It is important that the engineering program from which you graduate is accredited by the Accreditation Board for Engineering and Technology (ABET). It is always a sad day when a graduate of an engineering school learns that it was not an accredited program. Accredited schools can be found easily by going to the ABET Web site at *www.abet.org/accrediteac.asp*. It is also important to know that not every state will allow graduates from an engineering technology program to become licensed, even if the program is ABET accredited.

The FE (or EIT) exam is standardized and is produced by NCEES. It is the same in every state. The exam is given twice a year, in April and October. College seniors often take it because the April exam is given a month or so before graduation. There is no better time to take the exam than just before graduation or as soon as possible afterward. As you will see, the exam is very comprehensive and long. It is best managed by someone with fresh recollections of the subject matter and with a lot of recent test-taking experience.

The FE exam is closed book. Exam takers are furnished with officially sanctioned reference materials at the exam site, and no other reference materials are allowed. The exam takers will be provided with NCEES-furnished mechanical pencils and leads. There are strictly enforced limitations on the type of electronic equipment allowed into the room. The exams are closely monitored, and anyone detected to be violating any of the rules will be ejected.

Be absolutely certain that you can observe every rule and every element of the schedule. We have heard of latecomers not being allowed into the examination once the doors are closed, even if they are late by only a minute or two. Don't even think about breaking one of the rules.

The exam will be administered in two sessions of four hours each: one in the morning and one in the afternoon. In the morning session there will be 120 multiple choice questions, and in the afternoon session there will be

60 multiple choice questions in one of seven specialized areas to be selected by the exam taker. The subject matter of the exams will be as follows:

Morning session:
- Mathematics
- Engineering probability and statistics
- Chemistry
- Computers
- Ethical business practices
- Engineering economics
- Engineering mechanics (statics and kinematics)
- Strength of materials
- Material properties
- Fluid mechanics
- Electricity and magnetism
- Thermodynamics

Afternoon session:
- Chemical engineering
- Civil engineering
- Electrical engineering
- Environmental engineering
- Industrial engineering
- Mechanical engineering
- Other/general engineering

Clearly, the exams cover a vast array of topics. You really need to prepare to take the tests. There are many sources of preparation assistance. Kaplan AEC Education (*www.kaplanaecengineering.com*), a subsidiary of Kaplan Publishing Company, the publisher of this book, has review books for the FE and the PE exam. As of this writing, three state societies of the National Society of Professional Engineers (NSPE) have organized and conducted local review classes using the Kaplan materials. There are other resources as well. Information is available at the NCEES Web site about review materials, and classes may be given by schools in your area. However you do it, if you have been out of college for a while, you should give serous consideration to taking a refresher course before taking the exam.

One of the requirements to take the PE exam is four years of qualifying engineering experience acceptable to the registration board in your state. Most, if not all, want the experience to be in design, under the supervision of a licensed professional engineer. Keep that in mind as you choose your employment. Some states will not give credit for certain types of experience, specifically construction work. The Professional Engineers in Construction Division of NSPE has prepared a helpful document to help with making the most of construction experience. It can be downloaded from *www.nspe.org/ pec/Brochure2002.pdf.*

We would advise checking out what constitutes qualifying experience in any states in which you wish to become licensed. If the registration board is user friendly, you may be able to get information by simply calling or writing to the board. If not, you can ask your mentor at work or ask members of the technical or professional society to which you belong. You will need a specified number of references, with some or all of them being PEs. Keep that in mind as you spend your four years at work. Try to keep a list of those people with whom you have contact so that you can come up with the best candidates for your list at the right time.

Finally, after all the other mileposts have been successfully reached, you will be prepared for the last step, taking the PE exam. You will need to contact the registration board in the state or territory in which you wish to become licensed. You will be sent a package of material, including guidelines, application forms, and reference forms. The registration boards in each state or territory establish the precise criteria for eligibility. Once you are satisfied that you meet the requirements for the state or territory, you can then proceed to fill out an application for consideration as a professional engineer.

The application form will ask for your personal data, a photograph, and details about your education and work experience. Do not take any shortcuts when outlining your work experience. Remember, the board will be judging it for compliance with its standards for acceptability. Be sure you give a thorough and honest description of the work you have done, emphasizing the design aspects. Here is where a good mentor can give you some support. Ask for a critique of the application before you finalize it and send it in.

You will also be asked to have your college or university provide a transcript of your classes and grades, usually to be sent directly to the board. There will be a requirement for a specified number of references, both character and professional. Often the professional references are required to be provided by a PE with knowledge of the quality of your engineering experience, and the application may require the PE's comments be linked to specific periods of experience stated on your application. The board will expect you to be responsible for ascertaining that the transcripts and references have been sent.

After board review, you will be notified if you have qualified for examination or not. If you have qualified, you will be given instructions about where and when to take the exam. Some state boards will not do that themselves but may refer you to an outside organization, Engineering and Land Surveying Services (ELSES), to handle everything related to the examination. The ELSES Web site is at *www.els-examreg.org*.

The exams are given in April and October of each year. You will have to choose a particular category of engineering practice upon which you will be tested for both academic knowledge and knowledge gained through actual engineering practice. The engineering categories are:

- Agricultural
- Architectural
- Chemical
- Civil
- Control systems
- Electrical and computer
- Environmental
- Industrial
- Mechanical
- Metallurgical
- Mining and mineral
- Naval architecture and marine engineering
- Nuclear
- Petroleum
- Structural I
- Structural II

If a state-specific section is included in the exam, you will be informed in advance.

As stated earlier, study materials and courses are available to assist you in preparation for the exam. NCEES also has study materials available at its Web site. You can order them as you see fit.

For the civil, structural I, and structural II exams, certain design standards are required. You can find out about those standards at the NCEES Web site. You must bring your own copy of those design standards to the exam. NCEES cautions that if you refer to another standard, one not listed at its Web site, you may not get credit for the answer.

As with the FE exam, strict rules are associated with the process. Calculators are permitted but only those models listed by NCEES at its Web site. Cell phones and pagers are prohibited from the room. You cannot bring in your own pencils or pens. Only NCEES-sanctioned pencils, leads, and erasers are allowed. They will be furnished at the exam site. Other prohibited items include, but are not limited to, computers, calculators not on the approved list, and PDAs.

The PE exam is open book, but the books are subject to the rules of each state. The NCEES design standards are, of course, admissible, but remember, you must bring your own copies.

Some states require a special structural professional engineer license. The testing requirements for those states can only be determined through information provided by the respective registration boards.

Once you have completed and turned in your exam papers, you can relax for a while. The timing will vary, but you can expect to have results within three months of taking the test. The registration board will send your results to you. You will be notified only of pass or fail. You will not know your score. If you should fail, you will be furnished with a diagnostic report that will give you insight into your strengths and weaknesses as revealed by your exam performance. If you fail on the first try, please don't throw in the towel. The diagnostic report will give you important insights into what areas need additional work. Additional study and preparation in those areas will significantly increase your probability of passing on the next attempt.

Some states have rules limiting the number of attempts you can make at passing the exam without taking some supplemental steps. This, of course, is another area that can only be determined on a state-by-state basis.

There is a process whereby NCEES, for a fee, will hand score answer sheets to verify the outcome of any particular exam. This can be utilized within 60 days of the date on which NCEES furnishes the results to the registration board. The request can be made through the registration board or through ELSES, whichever is applicable.

NCEES determines a passing score for each exam as it is introduced for actual use. To do so, it relies upon a panel of professional engineers who are subject matter experts. The members of the panel are chosen for their expertise in specific areas and because they are deemed qualified to assess the ability level to be expected of applicants at a particular time in their career. The panel makes their assessment against a level that would constitute minimal competence, not anything higher than that. NCEES also uses a statistical process with the results of each exam to determine a passing level that will not penalize an applicant if the exam shows itself to be exceptionally difficult.

If NCEES determines that one of the exam questions was flawed, either by analyzing specific questions that result in unusually skewed statistical results, or by review precipitated by comments sent in by applicants, it will take some action to mitigate the consequences. It may go so far as to give credit for all answers given for that question.

We told you that the path to licensure would be time consuming and cost you some money. We also said that in the long run, the career enhancements would far outweigh the costs of time and money. What value does it hold for you?

- It is a clear indicator that you have personally accomplished something that takes a lot of time, effort, and cost to join a group charged with ethical conduct and a lofty set of standards.
- It places you a notch above others who have engineering degrees and maybe even experience in the profession but who don't have the professional resolve to prove it in a well-recognized way.
- It will allow you to be legally in charge of engineering work in the areas of engineering that are not exempt from the law.
- It will enhance your personal worth and value in the engineering marketplace.
- It will provide you with more options in the event you wish to (or have to) seek alternate employment.
- You will be a member of a group that is highly respected by the public.
- Your salary level is likely to be higher than that of an unlicensed engineering graduate in the same firm. Furthermore, your highest salary level will definitely be higher in most organizations. An exception to this statement can occur when your employer has a rigid salary scale structure that does not recognize licensure as a factor for a compensation increment. The *Wall Street Journal* has posted an article online at *www.careerjournal.com/salaryhiring/industries/engineers/20050822-engin-cert-tab.html.* The cited source of the data is the "Engineering

Income and Salary Survey," 2005, prepared by the National Society of Professional Engineers. The average annual incomes shown were:

- Professional engineer and other professional licensing: $105,875
- Professional engineer and surveyor or land surveyor: $107,093
- Professional engineer and certification in other engineering specialty: $96,114
- Professional engineer and certification in forensic engineering: $120,954
- Professional engineer and certification in environmental engineering: $99,482
- Professional engineer: $86,450
- Engineer-in-training/engineering intern: $56,480
- No professional registration or certificate: $72,911

It is always true that the average value in any group is not an absolute predictor of what your salary will be as a member of the group, but the message delivered by the data is inescapable. Being licensed is an asset toward higher compensation, and being additionally certified can be an even greater asset.

Let us close with a true story. A colleague of both authors was employed in industry for many years. She sought and attained a license as a professional engineer because of the urging of a superior in the company. It was not a requirement for her job, but it was a personal goal. Years later, she left industry and went to work in an unrelated field. She recently interviewed for a position in another unrelated field. The position had nothing to do with engineering, and licensure was not a relevant issue. She told us that she was chosen for the position, and the interviewer told her that the fact that she was a licensed professional engineer spoke volumes about her. That swayed the decision. She said that her salary on the new job was over $100,000 more than the salary she had received on her job in the original industry. Even though her new position is in an unrelated field, the PE license is a highly regarded designation that can help you achieve your professional goals, whatever they may be.

18

CONTINUING PROFESSIONAL COMPETENCY

As a soon-to-be graduate engineer, or one who has recently left the hallowed halls of education to enter practice, you might feel that you have all of the education you'll ever need to be a successful engineer. We know, however, that somewhere in the back of your mind is the reality that you still must learn lots of practical applications on the job—maybe even take an occasional course to stay current. In this chapter, we provide insights to engineering education and what you really need to plug into your future plans with respect to continuing education.

THEN AND NOW

When we graduated from our respective engineering schools about a half-century ago, studying engineering was quite difficult. Many of our colleagues failed to complete the curriculum in the targeted four years, which is not unlike the engineering programs of today. You must understand, though, that your education does not end when you receive your degree. Since we finished our academic studies, both of us have been on a continuous learning quest throughout our careers, with the greatest pressure to stay current with the advent of the microprocessor, the popularization of computer technology, and enhanced communication technologies. There have been enormous gains in knowledge about chemistry, materials, technology, communications, and so much more. In fact, many forces in the engineering community are now promoting conversion of engineering study to a five-year program. The ASCE, NSPE, NCEES, and others have studied the need for the "first professional degree" (the one necessary to attain licensure) to be a minimum of five years in length.

The body of knowledge necessary to produce a properly skilled and knowledgeable engineering graduate has dramatically expanded. Before we look at the issue of continuing professional competency, let's look at recent history. What we are about to review is important for you to understand with respect to the "first professional degree" issue. It can influence your future competitiveness in the professional marketplace.

Various professional and technical societies have focused their attention on the adequacy of engineering education in recent years. The National Council of Examiners for Engineers and Surveyors (NCEES) is the umbrella organization of registration boards in the states and territories of the United States. In 2001, NCEES formed an Engineering Licensure Qualifications Task Force (ELQTF), charging it with evaluating the licensure process in the United States. At the 2003 annual meeting of NCEES, the ELQTF reported that the national average academic load for engineering candidates had gradually declined to 128 semester hours. By comparison, in the 1950s I was required to attain at least 144 semester hours of educational credits to attain a BSCE. The ELQTF finding, of course, translated into less contact time for the transmittal of an ever-increasing body of knowledge to engineering students. The ELQTF reported the opinion that future engineering curricula would need to be expanded to provide adequate academic preparation for engineering graduates to enter the professional level of practice.

At about the same time, the American Society of Civil Engineers (ASCE) had produced a publication entitled *Civil Engineering Body of Knowledge for the 21st Century*. The ASCE postulated that the body of knowledge for the engineering profession was growing extremely rapidly, perhaps as much as doubling every decade. In line with that conclusion, the ASCE adopted Policy Statement 465, promoting additional education beyond the BSE as mandatory for licensure as a professional engineer.

Other organizations, such as the National Society of Professional Engineers (NSPE) has expressed similar concerns. In January 2002, the NSPE adopted Professional Policy No. 168, supporting additional education beyond the four year ABET/EAC degree as a prerequisite for licensure.

In 2004, the NCEES created a Licensure Qualifications Oversight Group (LQOG), charging it to further evaluate the conclusions of the ELQTF and to move forward with recommendations. The LQOG supported the conclusions of the ELQTF and found the following::

> The Body of Knowledge required for the practice of engineering in the future and for the continued adequate protection of the public health, safety, and welfare is beyond the scope of the current background provided in the traditional four-year engineering curricula in the United States.

The LQOG recommended that the NCEES Committee on Uniform Procedures and Legislative Guidelines (UPLG) should be charged with creating language to modify the NCEES Model Law and Model Regulations to require additional engineering education for licensure. The Model Rules and Model Regulations are not binding law but are the models by which most of the state and territorial legislatures and registration boards are guided in framing their own statutes and regulations. In general, most jurisdictions follow the NCEES models.

The LQPC recommendation was approved at the 2005 annual meeting of the NCEES. The UPLG committee prepared language to be used in the Model Law. At the 2006 annual business meeting of the National Council of Examiners for Engineering and Surveying (NCEES), delegates voted to modify the NCEES Model Law requirements for licensure to require additional education for engineering licensure. The approved language states that an engineer intern with a bachelor's degree must have an additional 30 credits of acceptable upper-level undergraduate or graduate coursework from approved providers in order to be admitted to the Principles and Practice of Engineering (PE) examination. The Council also passed a UPLG motion adding language to the Model Rules stating that, effective January 1, 2015, a graduate with a bachelor of science degree in engineering requiring more than 120 credit hours may request that hours earned in excess of 120 credits be applied to satisfy the requirement. The change will not affect any current engineering students.

LOOKING AHEAD

Keep in mind that in the not-too-distant future, you will face competition from a group of graduate engineers who have experienced a greater degree of academic training than you. To stay competitive with these new engineers, it would be prudent for you to consider your professional development options.

Should you take the advice that we have given throughout this book to become a licensed professional engineer, you need to be aware of continuing professional competency (CPC) as a legal requirement for the maintenance of an engineering license. Historically, CPC (also known as continuing professional education and continuing professional development) has come a long way in a relatively short time. In jurisdictions that require engineers to earn CPC credits for maintenance of licensure, the term becomes MCPC (mandatory continuing professional competency).

The first state to institute MCPC for engineering licensees was Iowa in 1979. Fourteen years later, a second state, Alabama, followed suit. Later on, in the 1990s, the movement toward requiring MCPC had accelerated. As of this writing, according to NCEES records, 29 jurisdictions have MCPC in place, and 5 more have indicated the expectation of having it within five years (by 2011). The jurisdictions with MCPC actively in place are:

Alabama	Mississippi	Oklahoma
Arkansas	Missouri	Oregon
Florida	Montana	South Carolina
Georgia	Nebraska	South Dakota
Illinois	Nevada	Tennessee
Iowa	New Hampshire	Texas
Kansas	New Mexico	Utah
Louisiana	New York	West Virginia
Maine	North Carolina	Wyoming
Minnesota	North Dakota	

The five that anticipate it by 2011 are:

- Alaska
- Hawaii
- Kentucky
- Ohio
- Pennsylvania

Other states are also studying the issue but with no indication of if or when it may be enacted.

In general, MCPC requires a licensee to obtain a specified number of professional development hours (PDHs) during each period between license renewals. A PDH is generally a 50-minute period spent in an approved activity. The number of PDHs may vary from one jurisdiction to the other, generally being on the order of magnitude of 12 to 15 per year. The requirements are established by each jurisdiction, although many comport with the NCEES Model Law for CPC. The character of acceptable learning experience will also be determined by the rules in each individual state. Certainly classroom time in technical courses will be acceptable. Sometimes classes in so-called "soft skills," such as engineering-related management or business courses, professional liability and risk management, or ethics are acceptable. Some jurisdictions will give credit for teaching time, for authoring articles, or even for being an officer in a professional or technical society or attending meetings. The exact nature of acceptable PDH activity must be researched on a state-by-state basis.

The NCEES Model Law makes a provision for you to meet the standards of any other jurisdiction's MCPC law if you meet those of your home jurisdiction. That means that if you work and are licensed in state A, you may be able to meet the MCPC requirements of state B by satisfying state A's requirements. That, too, must be determined by looking at state B's law and relations. Currently the NCEES and other organizations are trying to influence the various jurisdictions to become uniform in the rules associated with MCPC to facilitate portability of licensure, but the jury is out on the outcome of that effort. It cannot be overemphasized that the requirements of MCPC must be carefully researched for every state in which you may become or may be thinking of becoming licensed as a professional engineer.

THE NEED FOR LIFELONG LEARNING

Even if you do not intend to become licensed, or if you intend to become licensed in a jurisdiction that does not have MCPC as a requirement, you need to think seriously about the well-established reality that the engineering body of knowledge is explosively expanding. For your own professional development, as well as in the interest of maintaining or enhancing your marketability as an engineer, you absolutely must commit yourself to a sustained and robust program of lifelong learning.

You will undoubtedly learn engineering skills on the job in the mere performance of your duties, but that is not enough. Many will ardently argue the case, insisting that they can achieve the necessary learning in their day-to-day performance of engineering duties. However, legislators and registration boards in more than half our states have decided otherwise. The various committees within the NCEES have decided otherwise. The most prominent professional and technical engineering societies have decided otherwise. It is my personal opinion that those who argue for on-the-job achievement of professional development fail to see the big picture.

During my early career, I worked at a highway design firm. One of my supervisors said that in his 20 years of experience in highway design, he really had "5 years of highway design experience 4 times." By that, he was suggesting that there was not that much to learn in highway design—5 years' worth in his estimation—so 20 years just netted him 5 years of experience times 4. That was, of course, an exaggeration. What is true, however, is that the way an engineering firm utilizes your services may not afford you the opportunity to develop all of the skills that you may want to attain. You can take charge, however, and do it through continuing professional development courses taught by recognized experts. Not only that but the importance of sharing time with similar professionals who work in different areas of engineering cannot be overlooked. The collegiality that occurs during coffee breaks and during luncheons is invaluable to your professional growth.

You may be wondering about the cost of CPC and who covers it. First of all, there are some opportunities to obtain PDHs at no additional cost beyond dues if you belong to a professional or technical society. Sometimes, PDHs are offered at no cost or very low cost at society meetings. In addition, in some jurisdictions PDH credit can be earned by writing an article and having it published, and also for teaching a course in your field. Some states allow credit for attending society meetings and some for being a society officer. Of course, it is likely that some courses will require payment of a registration or tuition fee. Generally speaking, employers will reimburse you for all or some of the cost, provided they are satisfied with the subject material. Once all of those options are gone, then you may need to pay out of pocket. Generally, attending courses to obtain PDHs is not extremely expensive.

As a final thought, whether or not continuing education is required for your profession, you should still aspire to be a lifelong learner. If we, as engineers, want to demonstrate to the public that we are true professionals, just like doctors, lawyers, and CPAs, then we need to match the lifelong learning commitments made by those professions. MCPC exists for all of them in most states. REALTORS® are required to take courses, as are architects, landscape architects, land surveyors, and others. We, as engineers, have worked long and hard at attaining a position of status in the public eye. We need to rise to the occasion to maintain that image by conducting ourselves with the same dedication to lifelong learning as that exhibited by so many other professions. I sincerely hope you agree with this concept and that you will be guided by it as you move through a successful career in our wonderful profession.

19

PROFESSIONAL ASSOCIATIONS

Starting in college, you will be asked from time to time to join an association. You can find, in your library or on the Internet, long lists of professional associations for a large number of professions. Specifically for engineering, you can find a list of professional associations in the appendix and on the NSPE Web site, among other places. These are dynamic lists, because new associations are established on a regular basis. Additionally, some associations are discontinued or merged with other organizations because of the changing needs of the profession. Having said all this, the next logical questions are:

- What is the best association for me?
- Do I really need to get involved with an association?

DEFINITIONS

A *professional association* is defined as a formal organization of practitioners of a given profession. The association usually operates on a nonprofit basis. It exists to promote and develop a particular profession and in so doing to protect both the public interest and the interests of professionals.

Many professional associations perform professional certification or administer licensure to indicate that a person possesses qualifications in the subject area. Sometimes, membership in a professional body is synonymous with certification.

Also, membership in a professional body is sometimes required for one to be able to practice the profession legally. That is not true for engineering. Generally, there are two classes of engineering professional associations:

1. *Technical societies.* These include the American Society of Civil Engineers (ASCE), American Society of Mechanical Engineers (ASME), and the Institute of Electrical and Electronics Engineers (IEEE).
2. *Engineering associations.* These include the National Society of Professional Engineers (NSPE), Society of Automotive Engineers (SAE), the Association for Facilities Engineers (AFE), and the National Society of Black Engineers.

A *trade association,* which you may also want to consider for membership at some point in your career, is defined as a group founded and funded by corporations or practitioners that operate in a specific industry or trade. Usually, a trade association's purpose is to promote that industry or trade through public relations activities such as advertising, education, political activity, publishing, establishing standards, and collaboration between companies or trade practicioners. Engineers are most likely to join a trade association if their employer specializes in a particular product or family of products. Membership in trade associations of clients, such as the National Association of Home Builders, can provide excellent opportunities for marketing and for keeping abreast of industry news that may affect your practice.

LEGAL CONCERNS

Licensure of engineering professionals is usually administered by state government regulations. It is unusual for a state to assign a professional association of peers to provide self-governance in engineering and to be responsible for licensing practitioners in the field. It is more usual for the engineering association to assume a supporting role in regulating the profession.

The professional association exists to further a particular profession, protecting both the public interest and the interests of professionals. The balance between these two may be a matter of opinion. On the one hand, professional bodies may act to protect the public by maintaining and enforcing standards of training and ethics in their profession. On the other hand, they may be criticized for acting too strongly to protect the interests of the members of the profession. The engineering associations seem to handle this delicate balance well.

WHY JOIN?

Membership in an engineering association is a career-enhancing decision you must make. Because of limited time and resources, you may need to limit your membership to one or two associations. You need to perform a cost/benefit analysis on your membership to figure out if the benefits of the association are worth the investment of money and time. The membership section of many engineering association Web sites includes a template to help perform this cost/benefit analysis.

Let's look at some of the many reasons you should join an engineering association.

- *Information.* Newsletters, magazines, and e-mails all can keep you current in the latest developments and news in your field.
- *Networking.* Meetings, conventions, and online forums keep you in touch with your engineering peers, who may be able to help you with new ideas, special problems, career advancement, and job hunting.
- *Education.* Engineering associations have many courses, seminars, online instruction offerings, and links to many education opportunities.
- *Advocacy.* Your association actively represents and protects your engineering specialty in the legislative arena.
- *Public relations.* Your association may be a primary resource to the news media regarding your engineering specialty and may even provide a public relations program to promote it.
- *Career advancement.* Your association may provide listings of employment opportunities, résumé referral services, and occasional salary survey information.

- *Professional development.* I have met many engineers who believe that their participation in engineering associations enabled them to learn management and leadership skills more rapidly than they would had they relied on their employer. When you become active in an association, there are many opportunities for committee leadership, public speaking, team building, networking, greater understanding of related professions, and more. I believe that the social element of association meetings also enhances your professional development. Some associations schedule periodic meetings that involve spouses, which can go a long way to help your spouse or partner better understand your profession.

- *Products and services.* These include publications, standards, videos, document templates, and even paraphenalia with the association logo. More important, your association can provide insurance, health care, investment, retirement, and other fringe benefit services at cost-saving rates. As described in Chapter 16, the future workforce will involve more and more engineers working on a contract or self-employed basis. Engineering associations are the easiest way to provide fringe benefits at a group rate for the individual engineer.

WHO PAYS?

Engineering association membership requires time and money. Many engineers work long hours, and engineers with young, growing families have a limited amount of spare time. Thirty years ago, before the intense global competition we have today, employers could easily justify paying engineering association membership fees for their engineers. They would even provide time off with pay for their engineers to attend meetings and participate in association committee activities. Today, that fringe benefit is less common. That's one of the reasons that engineering associations have suffered declining memberships.

Unless your organization has a strongly worded policy that does not provide payment for professional society membership, you should discuss the concept with your supervisor. Before choosing an engineering association, ask your supervisor for advice on the best possible association for your career. Also, discuss with your peers which engineering associations they believe would be most beneficial to you. Also, you may have been a member of a student association for engineers; these associations usually have professional counterparts that should also be considered. Perform a careful investigation and analysis of all your options and make a proposal requesting your organization to pay for your membership.

If your organization will not pay for your membership, we recommend that you re-examine your family budget and make the necessary revisions to accommodate membership in at least one engineering association. Do not think of dues as an expense. Rather, think of dues as an investment in your personal professional development as well as a measure of support for your profession.

ENGINEERING STANDARDS

One of the roles of engineering associations is to establish and maintain engineering standards. Examples are the American Society for Testing and Materials (ASTM), the National Electrical Manufacturers Association (NEMA), and the IEEE, which establishes numerous electronics standards. Some of the standards issued by professional societies have the force of law if a legal controversy arises. As described in Chapter 12, failure to conform to specifications can have devastating costs if a product liability lawsuit is filed against your company and the plaintiff can argue that the injury or loss occurred as a result of your failure to conform to the code or standard. In the consulting world, professional negligence or malpractice suits will almost always look at the conduct of the engineer in the light of standards and codes relied upon by the engineer.

CREDENTIALS

Upon graduation, engineers possess the necessary academic credentials to enter the industry. Within a few years, hopefully, many also possess a professional license. Your involvement in an association adds another credential to your experience and education that other professionals within the industry highly respect. Once you've been a part of one of these associations, they often offer designations for life members, senior members, and distinctions for outstanding achievement. For instance, I am a Fellow Member of the NSPE, so FNSPE is one of my credentials.

In some associations within the design disciplines, such as the American Institute of Architects (AIA), members use this affiliation as they would an academic degree or license by including the AIA initials following their name. An association for executives is called the American Society of Association Executives (ASAE). Their credential, for those who earn it, is the Certified Association Executive (CAE). Some members of the ASAE are also engineers and use engineering credentials as well as CAE after their names.

With our global economy, you likely will work with engineers from other countries. If you do not recognize an affiliation, you should inquire about it, as some are not always what they appear to be. For example, engineers with only the equivalent of a bachelor's degree in most South American countries are permitted to use the title "Dr." in front of their name.

ARE ASSOCIATIONS DIMINISHING IN INFLUENCE AND MEMBERSHIP?

Unfortunately, association membership in general has declined in recent years. Young professionals today work long hours. They carefully balance career-related activities with quality family or hobby time. Consequently, more engineers today elect not to join professional associations or to join only as a "mailbox member." This term, used by association leaders, refers to members whose activity is limited to reading the newsletters or magazines. These members take information, and perhaps other services, but give little back to the association other than dues.

This trend in membership decline is unfortunate. We have seen that associations provide valuable services to their profession. Declining membership usually implies reduced services. Declining membership also implies declining legislative influence to protect the interests of the profession. Most professional associations are working very hard to reverse this membership decline. They use many tactics, but a key program is to list all the benefits and recent accomplishments of the association on their Web site. You are encouraged to review this list from time to time and seriously consider first becoming a member, then becoming an active member.

Membership in a professional association can be likened to membership in a health club. You can belong to a health club, paying dues faithfully because you know that the health club does good things in the community. But unless you actually go there, get involved, and participate, your personal physical well-being will not benefit. Your membership in one or more professional societies will be gratefully welcomed, of course, but unless you actively participate, your professional development is likely to fail to reach its full potential.

20

MENTORING

I stumbled into the engineering profession by sheer luck. A civil engineer and land surveyor at a state employment agency hired me the summer after my freshman year in college. I was enrolled at the time in the school of business administration. My field crew chief showed me some basic skills, and I spent about three months learning more and more from him. He quietly, without specific coaching, taught me about duty to an employer and about excellence in what you do. He, to this day, stands out as the reason I changed majors and became an engineer and land surveyor. I not only loved the work, but I wanted to be like him. What I most admired about him, and have tried to emulate, was his ability to teach me, a young, inexperienced surveyor, the right way to do things in a proper manner without making me feel like a dummy. He corrected my mistakes with respect, although he could also become stern if I consistently repeated the error. But he never did it in a demeaning way. He was also able to commend me for a good job. These are great leadership characteristics.

WHAT IS MENTORING?

Whether you are a college student or have recently entered the engineering profession, you need to be aware of mentoring: what it is and what it can do for you. You need not only to be aware of it, but you need to take a proactive role in seeking a mentoring relationship. A mentoring relationship, in simple terms, consists of a mentor and a protégé (sometimes called a "mentee"). A mentor can be defined as a trusted counselor or teacher, especially in occupational settings. A protégé can be defined as one whose welfare, training, or career is promoted by an influential person.

Probably mentors have already influenced your life. You may have had one or more teachers in elementary school or high school whose guidance is still a driving force in your life. Possibly members of the clergy, sports coaches, dance teachers, scout leaders, relatives, job supervisors, and more have mentored you about life, morality, ethics, diligence to work, belief in yourself, self-reliance, and so much more. Perhaps neither of you thought of your time together as mentoring, because it was simply a natural relationship between two people who respected and cared for each other. I can recall significant persons who served as my mentors and whose influence has guided me for decades.

Perhaps you have noticed that the introduction to mentoring above does not deal with the teaching of specific skills. For purposes of this chapter, the teaching of specific skills will be considered coaching or technical training. The focus on mentoring is more toward guidance in personal conduct. Allow me to share another personal mentoring experience I had with someone who is long gone from this earth but whose kindness in sharing his principles of conduct with me has vastly influenced my life.

About 40 years ago, I attended my first meeting of the local chapter of the NSPE. I entered a room full of engineers who were significantly older than I. One of them greeted me and made me welcome. As the evening went forward, there was a business meeting during which the members spoke about issues of significance to the engineering profession. The gentleman who greeted me, a new and uninformed member, emerged as an authoritative leader of the chapter. His guidance in my career began that night and lasted until he passed away a few years ago. His mentoring shaped many of my goals and aspirations. I remember him fondly. His influence on me has led me to participate as a worker and as a leader in my state and national engineering society. His influence has motivated me to pass on helpful guidance in engineering and professionalism to the younger engineers who are following my generation.

I will share one of my mentoring experiences with you. I frequently visit the Forum page of the NSPE Web site. There are many questions, most of which are answered by members of the organization. I was contacted offline by a young engineer who had an issue to discuss but did not want to do it publicly on the Forum page. She asked if I would work with her privately by e-mail. I told her that I would be delighted to try to help her work her way through her problem, which was related to her employment and her ambivalence about seeking employment elsewhere. We corresponded for several months. As her mentor, I never told her what to do. I just carried on a dialog and asked questions that prompted her to pursue other chains of thought. Eventually she made her decision and acted on it. During that period, we began to discuss other aspects of the engineering universe. She was so grateful to the NSPE for making the connection possible that she joined the society, and she has since served as a chapter president and worked on a number of state and national committees. Through the experience, we became good friends, and we still communicate periodically. In fact, she invited my wife and me to her wedding. That is one of my favorite mentoring memories.

TYPES OF MENTORING

Mentoring can take many forms. Casual mentoring occurs often when two people engage in discussion. For instance, you may have heard about a fellow worker's concern for balancing family time with time at work or with a hobby. He may have shared with you the principles that are helping to achieve a satisfactory balance. Another example might be a casual conversation with a project manager who shared the concept of duty owed to client, to employer, and to the public, and how she tries to balance those duties when they appear to be in conflict. There can be hundreds of examples, but those should be ample for you to grasp the picture. You, too, may have mentored others. Have you ever told an incoming freshman in engineering school about the need to balance the freedom of being away from home with the need to keep up with a difficult and time-consuming curriculum? Have you ever told a younger sibling about the hazards of smoking, drugs, drinking alcoholic beverages, and so on? These are examples of things that can be considered casual mentoring.

Another sort of remote mentoring has become readily accessible because of the Internet. At many Web sites, you can ask one or more questions on almost any topic. As I mentioned earlier, the National Society of Professional Engineers has a section of its Web site (*www.nspe.org*) known as

"Forums." One can post a question in several available topic areas and wait for someone to answer. The Web site is visited by a number of NSPE members, but there is no guarantee that every question will be addressed (although most are). Online mentoring relationships can be set up through organizations like MentorNet (*www.MentorNet.net*) that can become long-term, one-on-one relationships. These are useful tools that you can utilize in your professional and personal development. A list of some mentoring-related Web sites can be found at the end of this chapter.

In my opinion, the most effective kind of mentoring evolves from a one-on-one, person-to-person relationship that allows mentor and protégé to meet regularly and effectively. Ideally, the mentoring relationship for professional and career development will exist within the firm in which you work. While an effective mentoring relationship can exist without formal guidelines, it is worthwhile to consider some guiding principles. There are important considerations for both mentor and protégé. For this chapter, we will focus on those that are particularly important for the protégé to recognize and use for guidance.

FINDING A MENTOR

The first thing you will need to do is to make the decision to set mentoring as a goal. Once you have made the personal commitment to do that, you will need to find a mentor. If you are employed by a large, sophisticated firm, government agency, or industry, you may find that the organization has a formal mentoring program with rules, guidelines, and even a facilitator to help you get properly linked to a mentor. If you are an employee of a medium- or small-sized organization, mentoring may be less stringently organized. Whatever the situation, you will need to be proactive in letting someone know you desire a mentoring relationship.

It is usually best to link with a mentor who is working in the same organization as you, because often mentoring will include guidance on firm-specific issues. Should there be no availability of a mentor within the organization, however, you should not give up the quest. You may be able to find a mentor from an outside organization, one you hope is similar in practice. If you are a member of a professional or technical society, such as NSPE, ASCE, ASME, or IEEE, you will have an excellent opportunity through networking to seek out a willing mentor.

A mentoring relationship is not like a parent-child relationship, although good parents do, of course, mentor their children. But parents have a much broader scope of relationship with children than mentors will have with protégés. Parents, especially in the earlier years of your life, had the authority and privilege of making decisions for you, of directing you firmly in your actions, and, of course, of exercising disciplinary control over you. A good mentor will do none of those things.

Once you have committed to seeking a mentoring relationship and have found a suitable mentor, you need to take some additional action to prepare for a fruitful mentoring relationship. The following list is by no means complete, nor will we explore the tasks in depth, but it is a good starting point that gives you a general outline of important steps you need to take to make the relationship successful.

- *Set up a general schedule of mentoring sessions.* The agreed-upon schedule should include the frequency of the sessions, the amount of time that you and your mentor will commit to each session, where you will meet and for how long a period you will commit to sustaining the mentoring relationship, at least for a first round. You may also want to discuss "plan B" procedures for times when the inevitable conflicts arise to disrupt a scheduled session.

- *Make a solemn and mutual commitment regarding confidentiality.* Each of you needs to trust the other and to feel comfortable with expressing open and candid feelings and thoughts that should never leave the room.

- *Discuss the possibility that situations may arise between scheduled sessions.* At such times, you will want to have some time with your mentor. However, you must respect your mentor's schedule. Extra sessions must be kept to a minimum and must absolutely be at the mentor's convenience.

- *Attend your scheduled sessions prepared to participate fully.* If you have a particular issue that needs discussion, be prepared to present it fully but succinctly. Do not hold back information that you think may put you in a less favorable light. Be open and honest.

- *Be open to your mentor's guidance.* Obviously, you need to have faith in the maturity of your mentor. Hopefully, you will have dispelled any doubt at the outset of the relationship. In a good mentoring relationship, your mentor will not be directive and will not tell you what to do. Guidance may come in the form of your mentor's recounting some similar experience(s) from his or her early career, or it may come in the form of a series of questions intended to help you reason your way toward resolution of your issue, toward setting a course of action for yourself.

- *Follow through on action plans set at your mentoring sessions.* Be prepared to report your progress at the next session. If conditions arise that deter you from following through on your course of action, be prepared to discuss that with your mentor. That may lead to further discussion toward resolving your issue in some other way.

Mentoring is an experience that will give great personal satisfaction to both mentor and protégé. For you, the protégé, the experience can go a long way toward shaping your personal behavior on the job and in the community for years to come.

ADDITIONAL RESOURCES

It would be advisable for you to look beyond the content of this chapter. Literature is available with more detailed information. One very good resource is the "Mentoring Guide for Small, Medium, and Large Firms," a product prepared by the NSPE Professional Engineers in Private Practice Young Engineers Advisory Council (YEAC). It is available for download for a small fee (free to members) at the NSPE Web site; look in the catalog of products. I was acting as advisor to the YEAC during the preparation of the manual and know firsthand how diligently and extensively the group labored to prepare an excellent resource.

Following are additional resources to help you with your mentoring program:

- Society of American Military Engineers: *www.same.org*
- What's Hot—Top Mentor Publications: *www.peer.ca/topmenbks.html*
- International Telementor Program: *www.telementor.org*
- International Mentoring Associations: *www.wmich.edu/conferences/mentoring*
- Mentor Centor: *www.mentor-center.org/icem.cfm*
- Mentoring for Business: *www.e-mentoring.com*

- NSPE Mentoring Programs: *www.nspe.org/pdlmentoring.asp*
- Personal Strengths Publishing: *www.personalstrengths.com*
- ORA Personality Profiler: *www.oraonline.com/renewal/html/software.html*
- National Council of Examiners for Engineering and Surveying: *www.ncees.org*
- National Society of Black Engineers: *www.nsbe.org*
- Society of Hispanic Professional Engineers: *www.shpe.org*
- Center for the Advancement of Hispanics in Science and Engineering Education: *www.cahsee.org*
- Society for Women Engineers: *www.swe.org*
- National Association of Women in Construction: *www.nawic.org*
- MentorNet (an e-mentoring network for women in engineering and science): *www.mentornet.net*
- National Action Council for Minorities in Engineering: *www.nacme.org*

21

FINDING
ANOTHER JOB

Adecade or so ago, anyone who frequently changed jobs was often considered an employment risk or even an unstable person. Today, if your résumé does not show several job changes, you are considered to be just out of college or not very aggressive. The old quest for job security, employment for life, and company loyalty is passé. But don't change jobs for the sake of job hopping or yield to the "grass is greener on the other side" syndrome. Job changes must be very carefully planned and executed. This is true even if you are forced into it by downsizing, mergers, outsourcing, or even departmental failure. We hope these issues will not arise for you, but you at least need to understand the best methods to look for another job in your quest for career advancement.

In other chapters, we discussed résumés and interviewing techniques. We will leave the mechanics of external job search to other experts and sources. A good resource is *www.monster.com* or *www.hotjobs.com*. Here we will discuss the politics and ethics of job searching.

USING THE
INTERNAL JOB-POSTING SYSTEM

Most medium- to large-sized organizations have an internal job-posting system. Such systems exist because filling vacancies from within, thereby reducing training and orientation costs, is usually less expensive than hiring from outside. These jobs are posted on bulletin boards, the company newsletter, or in the company computer network. Make sure that you read the postings on a regular basis; they are an opportunity to make a nice jump in your career progress.

Occasionally, internal vacancies may not be posted because management wants to bring in some outside expertise or seeks "fresh thinking" in a department or division.

Some internal job-posting systems are very formal and are directly linked to employee development programs. In these more formal posting systems, the employees are actively encouraged to review job vacancies regularly. The formal job-posting system often exists to fulfill the requirements of the Equal Opportunity Employment Act. It sometimes happens that a position is posted even though the hiring manager already has candidate in mind. EEO procedures require a job posting to give other possible candidates an opportunity for interviews. If you suspect that a candidate has been preselected, discretely ask a few people about this possibility. If the answer is yes, then you may wish to pass on the opening and save the work required to prepare for the interview.

In a very small organization, there may be no internal job-posting system. Here the loss of an employee from one internal project to another internal project may cause serious interruptions and project delays. While management may desire to make promotional opportunities available to everyone, sometimes it simply cannot afford to do so. Also, it could be argued that in a small organization, management knows all the employees and can communicate about openings as well as fill openings on an informal basis.

LESSONS FROM EMPLOYMENT ADS

Former New York City mayor Ed Koch, who constantly sought feedback from his constituents, had a famous slogan: "How am I doing?" You need to periodically ask yourself that question regarding your career. To answer the question, you need to check your progress against benchmarks. A good source of benchmark information is the employment ads in newspapers and professional journals. First, you can learn about which companies or which departments are hiring and which are not. Look specifically for activity at competitor companies. What technical specialties seem to be in greatest demand? What products seem to be hot and on fast-track development? Do the ads identify the type and level of engineer desired? Do they identify the pay level? How does your current status compare?

What do you do if you find a great-looking job opportunity? Go for it! Send a résumé or make a phone call. If there is mutual interest, go for an interview. You just might discover a better opportunity. Even if it does not work out, it keeps your interviewing skills sharp. And you may find out that your current job and salary are not so bad after all.

Returning to the question, "How am I doing?" can any other benchmarks provide answers? Yes. Several engineering associations publish salary surveys. Occasionally, you can find summaries of these surveys published in the association's newsletter or magazine.

RELATIONS WITH YOUR SUPERVISOR

When looking for another position, how do you play the cards with your current supervisor? You should not fear that your supervisor will retaliate against you somehow when he learns that you are looking elsewhere. A good supervisor would certainly not like to lose you but should not object to your quest for career improvement. If your supervisor cannot react in this fashion, then it is appropriate that you are looking elsewhere.

Now is the time for you to have a talk with your supervisor to maintain a smooth relationship. Tell your supervisor that you are happy with your current position (even if that is not 100 percent true), but you are always open to improving your career. Then, if you find a better position elsewhere, that makes your supervisor look quite successful as a developer of employees.

Upon hearing that you are considering another job, your supervisor may indicate that your opportunities are limited in your current position—this means you have confirmed your suspicions. Or your supervisor may indicate how valuable you are and encourage you to stay. That is a good response.

Now you can work with your supervisor to develop a plan for your next raise or promotion.

Keep the fact that you are looking for opportunities elsewhere to yourself as much as possible. If co-workers learn of your activity, you may be treated as a non–team player. If too many other individuals in the organization learn that you are looking elsewhere, they may develop a negative impression of your supervisor.

When an interviewer asks if your supervisor knows that you are interviewing, your best reply is: "Yes, and she does not want to stand in the way of better opportunities for me." If you are in a dead-end position or have a deteriorating situation with your supervisor, you must explain that as well as you can. Also, be aware that if you have a deteriorating situation with your supervisor and the hiring manager calls for a reference check, your supervisor has two alternatives:

1. In the interests of getting rid of you, your supervisor will give a glowing report.
2. If your supervisor is vindictive and wants to continue to make your life uncomfortable, your supervisor will give a negative report. If this happens, you need to again meet with your supervisor and try to repair the relationship.

Real "head games" can be at play here. Engineers usually prefer dealing with facts, figures, and formulas—not head games. But this part of your career development requires you to pay attention to the psychology of the situation. Good luck!

NEW POSITION: INTERNAL VERSUS EXTERNAL

If you are successful and get an offer from another company with a 20 percent raise and an opportunity for more rapid advancement, acceptance is a no-brainer, right? Don't be too hasty; there are other things to consider:

- Is there a decrease in vacation days?
- Is there a loss or decrease of retirement benefits?

- Is there a decrease in medical coverage?
- Is there a decrease in seniority?
- What are your out-of-pocket relocation costs?
- Will your spouse or partner need to find other employment?
- Will there be an increase in local cost-of-living expense?
- Will there be an increase in commuting costs?

Consider all these factors. Consider the counteroffer, if any, from your present employer. Then make an intelligent choice. If you decide to go to another company, be sure to leave on good terms. Thank your supervisor for all his efforts that contributed to your development. Thank all your co-workers for the opportunity to work and learn with them. Remember, you may need occasional help from them in the future.

ETHICS OF JOB SEARCHING

It seems that everyone is working long hours today. How do you find time to look for a better job? The easy answer is to search on company time. One survey indicates that one-quarter of workers who use a computer at work admit to searching on company time. It has been observed that job site traffic spikes on weekdays during lunchtime hours.

You are not being paid to look for jobs elsewhere, however, unless you were specifically requested to do so. Therefore, it is not ethical to conduct job searches on company time. However, it *may* be acceptable to conduct a job search using company-owned equipment during your lunch break. Know the rules; why jeopardize your current job? Furthermore, it is not smart to send your résumé from an e-mail address that identifies your current employer. You don't want to send the message to a potential new employer that you conduct job searches on company time.

The best alternative is to adjust your personal schedule and conduct your job search on your own time. Devote yourself fully to your job in the hours you're there, and job search with a vengeance in the hours when you're not. That way, you'll continue to get the praise and recognition to keep you on track at your current job. Don't shoot yourself in the foot by making co-workers suspicious and then maybe not landing a new job after all.

TRANSITION TO THE NEW JOB

If you are successful in your job search and you decide to accept the position in another organization, how do you make an effective transition and optimize your chances for success?

Your first few weeks on the new job create a strong impression. As long as you show intelligence, versatility, and a willingness to work and learn, people will be happy to have you aboard and will root for you to succeed.

Do some homework prior to day one on the new job. Check your interview notes. Commit to memory the names and titles of everyone you met and interviewed with so you will be able to greet them and pronounce their names correctly. Review the company's newsletters, annual reports, and press clippings. Know where and when to report on day one and get there early. Notice the schedules and work habits of coworkers so that you'll know the optimal times and means to connect with others. Look and act as if you're happy to be joining the team.

Start to keep a journal of procedures. Record names of key people and contact numbers. Be kind and appreciative to everyone who helps you learn the ropes.

Expect and embrace the inevitable challenges of your position. Maintain a flexible attitude; this will decrease stress for you and others.

Observe your boss's personality and work style, and tailor your interactions to her preferences. Make sure that you and your boss are on the same page. Find out

- what priorities and issues need to be immediately addressed
- how often and in what format you should provide project or work updates
- how your performance will be evaluated

Get to know as many coworkers as you can. Establish the foundation for a relationship, and trust and information will follow. Find out who the decision makers, influencers, stars, and up-and-comers are. Notice the traits they have in common and try to emulate them.

Listen more than you talk. Resist offering opinions or assessments before they are requested. You'll get more respect by listening and absorbing what your coworkers have to teach rather than by showing off how much you know.

Show lots of initiative. When you finish assignments, ask for more. Pick projects that have support from upper management and buy-in from co-workers. Be a team player. Don't engage in gossip. Always make your boss look good. Share credit with your workmates.

22

OVERSEAS ASSIGNMENT

We all love to travel. Some engineers combine their love of travel with the ability to increase their income by taking an overseas engineering assignment. It sounds intriguing and exciting, but working in a foreign country is not for everyone. What you see of a foreign country as a tourist is quite different from "living on the economy" of that country. I have experienced this phenomenon, and I believe that it is a marvelous learning experience—and sometimes a shocking experience. There are many advantages and many disadvantages of working abroad, and these vary greatly by country.

Most of the four million nonmilitary U.S. citizens living and working abroad enjoy their experiences, in part because they planned and managed the assignment well. Working abroad is much more complicated than moving your desk and your books to another country and continuing to solve engineering problems. It involves assimilating to a new culture, learning a new language, dealing with a new set of laws and regulations, learning and applying new engineering standards and codes, and effectively coping with a new set of both professional and personal challenges.

This chapter is about overseas assignments that require temporary relocation. However, much of this material can also apply to shorter-term assignments of a few weeks.

TO GO OR NOT TO GO

Accepting a new assignment is exciting. It usually involves career growth, greater rewards, new challenges, and new opportunities. An overseas assignment is all this and more. An overseas assignment enhances your value, because American business and the practice of engineering are becoming more globally oriented. The exposure to other cultures, customs, ideas, viewpoints, and work practices provides invaluable experience to you and to your family. An overseas assignment is also a valuable addition to your résumé.

Your acceptance of such an assignment must depend on full support and proper planning by your employer. Your employing organization must, first, be sure that your assignment meets a clearly defined business strategy. There needs to be a position analysis and clear job description based on bicultural input from both the domestic and foreign groups. This position analysis and job description should outline performance expectations for the assignment and become a part of your development plan.

Once overseas, your performance appraisal can be linked directly to the job description and development plan, thus laying the groundwork for a successful assignment. This process also helps ease your repatriation so that a suitable position is waiting for you upon your return, letting you stay on career track.

The decision to work overseas is very personal and must involve a review of the personal characteristics of both you and your family. You must, of course, meet all the educational, technical, and experiential requirements for the assignment, but you should also meet these personal characteristic requirements. The engineer must

- be flexible in response to ideas, beliefs, or points of view
- have respect for others and be attentive and concerned in a way that makes others feel valued
- be a good listener who accurately perceives the needs and feelings of others
- have a demonstrated ability to build and maintain relationships with people of diverse backgrounds
- be calm and in full control when confronted by interpersonal conflict or stress
- be sensitive to local realities, whether social, political, or cultural
- have a strong sense of self and be proactive, frank, and open in dealing with others

- be open, outgoing, and nonethnocentric
- be self-confident with a strong desire to take the initiative
- have harmonious family, personal, and professional relationships

As you perform a careful self-evaluation based on these requirements, you should be realistic about the constraints and barriers to effective performance but remain fairly optimistic about success.

If your overseas assignment is from within your current organization, you may be asked to take some personality tests or you may be sent to an assessment consultant to ensure that you are suitable for the assignment.

Your decision to accept the position requires full support from your immediate and extended families. If you have a reluctant spouse or dependent, aging or ill parents, or teenage children who must stay behind, you are at risk for an unsuccessful assignment. If you have any special needs such as personal or family health problems or special educational requirements for your children, you are at risk for an unsuccessful assignment.

Some of the above selection criteria are flexible, depending on the "foreignness" of the host country. If you are assigned to work in London where language and social customs pose few obstacles, you may experience little difficulty. On the other hand, if you are assigned to Dubai you will need a great tolerance for ambiguity.

To make an overseas assignment fulfilling and successful depends on how well you use the opportunity. You need to review your current career status carefully, possibly with the help of your mentor, and consider how an overseas assignment can help your career growth. You need to review your employer's culture and status in the international marketplace. You need to determine how marketable or transferable potential new knowledge and skills will be. Interestingly, many CEOs of large corporations have an overseas assignment in their résumé. If you carefully perform all these evaluations, you will make the decision that's right for you.

THE DOWNSIDE

Reports by those returning from overseas assignments are often like reports by those returning from Las Vegas—you only hear about the winnings. Many returning expatriates have a positive experience and love to talk about it. Those with a negative experience tend not to want to discuss it.

In the United States, we enjoy a very high standard of living, and we enjoy among the best living conditions in the world. In other countries, you may have to cope with extremes of climate, unfamiliar religious restrictions, political and legal restrictions, sometimes unsanitary conditions, high pollution levels, significant health risks, and greater environmental problems. These situations are exacerbated in third world countries. And in third world countries, where poverty levels and unemployment levels are very high, you must cope with other issues such as high crime levels, security concerns, and political instability. Also, many other countries do not have effective controls for political corruption. These are challenges; they make life interesting, and they offer invaluable lessons.

FACTORS FOR SUCCESS: THE DEVIL IS IN THE DETAILS

This section describes some of the issues you need to consider and resolve before embarking on an overseas assignment. It discusses both preparation and things you must do when living in another country. Proper preparation is much more than packing your suitcases and updating your passport.

Assimilate to the new culture. Adaptation to life in another country requires flexibility and accommodation skills, as well as a desire to experience new things and a willingness to take risks. Company human resource professionals commonly think that once employees have been trained and indoctrinated, they have learned the skills to be effective in the new environment. In fact, the learning process never stops. The ability to work well in another culture is developed over time through a complex process of acculturation that cannot be learned in a classroom. Success in overseas assignments appears to have much more to do with emotional issues than with IQ. Key emotional issues are the ability to perceive and manage social relationships effectively through self-awareness, self-regulation, motivation, empathy, and social skills.

Learn the language. American managers and engineers sometimes believe that, because associates in other countries speak English, there is no reason to learn another language. This may be true for business meetings and negotiations where translators are often provided. However, only through learning the local language can one truly understand the culture and hence comprehend the many forces affecting the local business environment. You can operate much more effectively on the local economy if you learn the basic necessities of the local language. Also, the natives will respect you much more if you show an honest effort to learn their language—they will not consider you to be just one more "Ugly American."

If you think that the above items are not really necessary, consider the fact that Europe and the United States are struggling with problems of large-scale immigration. Some of the most frequent criticisms of immigrant populations are a perceived refusal to assimilate to our culture and learn our language.

Cultural sensitivity. There is no substitute for cultural sensitivity. Read about the culture before you leave. Read about the history of the country before you leave. Learn and respect the nuances of everyday cultural behavior. Learn the religious restrictions and adopt the 11th commandment: Thou shalt not proselytize for your faith. If possible, talk to individuals who have lived there. While you were hired to do an engineering job, your effectiveness will depend on your ability to work within the context of the culture. Look for opportunities to make learning a two-way exchange and express your appreciation of the people and their culture.

Financial considerations. You need to research the sometimes complex financial issues specific to working in another country. These include reimbursement for cost-of-living differentials, reimbursement for travel and relocation expenses, reimbursement for, and frequency of, return trips home, reimbursement for local taxes, provision of adequate insurance on your life and your property, and reimbursement for expenses unique to living in the foreign country. You may need to consult with international tax experts. Be sure to get all of these issues resolved in writing before you accept the assignment.

Your family. Many people cannot tolerate separation from family for extended periods and opt to take the family along. This introduces an order of magnitude of complexity to logistics and support. You need to locate appropriate schooling and adequate medical care. You need to work out medical insurance issues. The overseas assignment will be enriching for the entire family if you are allowed to maintain your balance between family and work. Your children will assimilate to the new culture and language much faster than you do; it is a rich experience for them. By considering all issues and by careful planning, you will enhance the value gained from sharing an international adventure together.

Corporate support. In the Workplace of the Future section of Chapter 7, we discuss how modern technology has made this world a smaller place and made operating with multinational project teams easy. Despite this fact, the home office can all too easily forget about you in that other country. Real-time response may not be easy, so the work dynamic may change dramatically. In theory, everyone works together to ensure project success. However, in reality, working together may involve time delays. It may also be difficult to get timely communications, documents, prints, and news from your home office. It may be up to you to develop solutions for this problem or to plan work-around techniques to avoid any delays.

Your company also needs to provide any necessary legal assistance as well as support to ensure engineering compliance with local laws and regulations. This includes support for approval and sealing of engineering design work in accordance with local regulations. Obviously these regulations vary extensively from country to country. A good resource for information regarding these issues is the U.S. Council for International Engineering Practice (*www.usciep.org*).

SUMMARY

Engineers on assignment overseas will always have to deal with issues related to culture, language, social and local customs, and foreign infrastructure. You need to be flexible, adaptable, and patient. And you will almost always have to forego some of the comforts and amenities found at home in favor of enjoying those found in other countries. All of the inconveniences, however, are more than offset by the overwhelming positives.

Remember the Boy Scout motto: Be Prepared. Expect the unexpected. If you properly plan and anticipate a positive experience, your overseas assignment can be among the most rewarding of your career opportunities.

SOURCES OF INFORMATION

- C. N. Weller, Jr., *The Overseas Assignment: A Professional's Guide for Working in Developing Countries,* (Tulsa, Okla: PennWell Corp., 2006).
- U.S. Department of State (*www.state.gov/travelandbusiness*)
- Your human resources department
- Others who have worked there

INDIVIDUAL AND TEAM PROJECT CONCERNS

23

THE ENGINEERING
DESIGN PROCESS

During Engineers Week, I meet with 3rd and 4th graders to teach them about engineering design. When I visit the students, I am interested in introducing them to the idea that we engineers perform very important services to humanity, touching everyone's life in many ways on a daily basis. To do that, I need to be sure that they understand what I mean when I say *design*. I explain that *design* means a process used by engineers to solve problems, thereby filling people's needs. I explain that someone will have some sort of problem or need and that he may engage an engineer to help solve it.

To solve it, I tell them, the engineer will have to discuss it with the client to understand the problem fully, and then the engineer will think about it for a while, reflecting on how others have solved similar problems. The engineer will make a few sketches, maybe think about how much implementing the different ideas may cost, then discuss the ideas with the client. Then, I tell them, the client and the engineer will come to a conclusion about which idea will be most acceptable to the client, and the engineer will prepare plans in great detail showing a builder how to construct the solution, to turn an idea into reality.

Basically, I want to leave the students with the concept of engineering being a problem-solving profession, with the results usually being something that will make someone's life better, simpler, and safer. I finish that segment by suggesting a hypothetical problem. I tell them that I need to cross a river and ask for ideas. I get suggestions like "swim across," "ride a horse through the river," "rent a canoe," "build a raft," and, of course, "build a bridge." I conclude with praise for their engineering thought process, then explain that the best solution, the bridge, will last for a long time but will also be the most expensive, and that it will need to be designed by an engineer for others to build. They seem to grasp the concept quite well, as evidenced by the sweet notes and letters they usually send me after the session.

DESIGN PROCESS MODELS

Although a simple exercise conducted with elementary school students, the above discussion is parallel to the formal engineering design process. Many models are available in the literature. The simplest is a five-step process that 3rd graders readily grasp:

1. Attain a clear definition of the problem or need to be addressed
2. Conduct preliminary research and data collection from a myriad of sources
3. Conceptualize multiple solutions to address the problem or need
4. Review the various concepts, analyzing and evaluating them against a variety of criteria (cost, time to complete, resources and support systems available, environmental constraints, etc.)
5. Choose and implement the concept that is most desirable

Other models abound, however, depending upon where you look. Most models seem to focus on the practice of engineering in the manufacturing arena, including elements for building and testing prototypes before moving into the full-scale manufacturing process. This process gives the manufacturer and the engineer time to seek out latent defects or even test the marketability of the item designed. That element of the design process is very effective but not readily practicable in the civil engineering field, where the designs are often of large, unique facilities, such as bridges, dams, airports, sports stadiums, sewer and water distribution systems, sewage and water treatment plants, highways, buildings, and more. Fortunately, one can build and test scale models for large projects and conduct computer-modeling simulations of natural forces that can affect them.

Many designs have failed dramatically—but might not have if model tests or computer simulation had existed at the time. In the civil engineering world, a historical calamity occurred in the state of Washington in 1940.

The Tacoma Narrows Suspension Bridge opened for traffic in July of 1940. The bridge displayed instability in heavy winds. It gained the nickname "Galloping Gertie" because of its gyrations. On November 7, 1940, Galloping Gertie gyrated so violently in a heavy wind that the bridge virtually disintegrated, most of it falling into the water below. It was a singular disaster. You can see photos of the bridge and its failure easily by doing an Internet search for "Tacoma Narrows Bridge" or "Galloping Gertie." Since that disaster, the engineering design process for suspension bridges has included wind tunnel studies using scale models of the design. Obviously full-scale models of bridges, tunnels, dams, buildings, and the like cannot be constructed just for testing. We can be thankful for the engineering design process that has developed methods for model testing and for the creation of computer programs that allow realistic simulations and nondestructive testing of engineering designs.

Let's take a look at a typical engineering process from the manufacturing perspective. Then we'll follow up with a brief look at a civil engineering project as well. We will use a few more steps in our hypothetical project than is usual in the conventional five-step model.

THE MANUFACTURING PERSPECTIVE

The first step in the process is to ascertain what the problem is and who needs it solved. That may be your employer—a manufacturer of household appliances—but the real person in need is the manufacturer's ultimate consumer, the homemakers of the world. Let's assume a hypothetical situation in which you are asked to look into developing enhanced and simpler controls for an existing appliance currently manufactured by your employer. Your first step would be to gain a clear definition of the problem, or need. At a minimum, you need to explore the current state of controls for similar appliances around the industry to establish a benchmark for improvement. Bear in mind, however, that you should not be looking into attaining information that would be categorized as trade secrets. You will also want to talk with customer service in your own company to see if it has a compilation of customer complaints relative to the appliance of interest. You may be able to hold focus groups of consumers to learn what they would like to see, and you may be able to conduct some consumer surveys as well. You will need corporate approval to gather information this way.

Once you have gathered all the information you can reasonably attain, you need to generate ideas to lead you further into the process. This is the ideal time to form a design team to participate in the development of a broad list of functional improvements that might enhance the appliance. The design team will do that based upon the data that you collected in the first step. When forming a design team, it is important to determine the skill sets that need to be brought to the table.

When the design team has been acclimated to the project goal and has had the benefit of the information already collected, it is time to go into a brainstorming session. Everyone will be asked to look at the various needs to be addressed and then to throw out ideas toward solving the problem. The process works best if everyone on the team feels comfortable with contributing. It must be made very clear that, at this stage, no idea is a bad one, nor is any idea too remote to be put on the table. This is a time to abandon engineering practicality and let imaginations run wild. Flip charts can be used to record all the ideas thrown out. Keep reiterating that everyone should participate. This is a conceptual exercise, and it should not get bogged down with detailed debates about why any idea is not good or how to actually implement an idea.

When all ideas have been recorded and put into some form of document for distribution to the team members, it is time to move into evaluating the various proposals in relation to the perceived problems and needs determined at the outset. To address those needs, you will have developed desired outcomes of the process. Some of those outcomes may be broken down into subparts by the team. The initial outcome set by our hypothetical project was to develop an existing major appliance into a model with enhanced and simpler controls. As a result of your research, you gained more detailed information about what consumers might want. Perhaps some of the subparts of the major outcome might be

- more reliable temperature control;
- easier way to preset timer controls;
- battery backup for clock and other controls in the event of power failures;
- reduction in frequency of product failure; and
- improved and easier-to-read user manual.

With the outcomes and outcome subparts on the table, the team should revisit the long list of proposed solutions with an eye toward eliminating the least promising ideas as they relate to the outcomes list. Those that remain after that exercise should be looked at for overlap and for potential linkages or merging. All of this must be done methodically (back to engineering thinking). The techniques to guide this process are beyond the scope of this book but have been published elsewhere.

Once the list of surviving proposals is finalized, you will begin an assessment phase to determine feasibility. The team will now move well beyond the free form of the brainstorming session and will be as analytical as possible. In some instances, the team may decide that prototypes need to be constructed for evaluation. Perhaps it will be possible to retrofit an existing model of the appliance under study, for example if newer-generation electronic controls, switches, or connectors might be substituted for existing components. Some modeling techniques use mathematical modeling or even computer simulations. Some of this you may already have been taught in college. Literature is available as well. For the purposes of this book, we will just acknowledge that such modeling capabilities exist.

Once you have cleared the assessment of feasibility stage for the various proposals, you are likely to reduce the list significantly. It will then be time to perform a more detailed analysis of each surviving proposal. Again, the modeling process should be your main technique for evaluation. Whether you use mathematical modeling, employ computer simulation, or build a prototype, you will need to evaluate measurable results objectively with respect to each proposal's model. You should engage others in the process to aid in objectivity by adding their perspectives. The results must be analyzed to see if the desired goals are met and also to see if any shortfalls or deficiencies in the models are readily correctable.

When that phase is satisfactorily concluded, it is time to present your results and recommendations to your client and/or to the management of your organization. Here is where you will be called upon to use those good communication skills your professors kept harping on.

Consider your audience. Who will be there? What will they want to know? What level of technical language will they want to hear? To the extent possible, you should tailor your presentation to suit the audience. If you will be reporting to a senior level engineering manager, you can feel quite comfortable in using engineering terminology, discussing the technical factors of your process to date, and explaining why they led to the recommendations you are submitting. If, however, you are presenting to the corporate president who has an MBA, you must assume that she has virtually no engineering expertise. Your presentation may refer to the general processes you followed but should not be full of technical jargon that will bewilder and bore the listener. By all means, you should bring the entire engineering backup that supports your presentation, but it's not wise to overwhelm your listener with details that he won't comprehend.

As a forensic engineer who needs to testify to juries of laypersons, I have learned to avoid discussing engineering formulas, instead finding analogies to engineering phenomena that occur in the everyday life of ordinary people. My objective is to inform the jurors, not to impress them with my mastery of engineering.

At this point in our hypothetical project, management will either accept or reject your recommendation or refer it back for more study. If approved, the project will move forward for additional finalization of design, if necessary, and then move into manufacturing and sales. You will have used your hard-earned engineering degree and knowledge to improve an appliance that will find its way into homes, where it will give years of enhanced service to countless consumers.

THE CIVIL ENGINEERING PERSPECTIVE

At the outset, I pointed out that in the civil engineering sector, it would not be possible to parallel exactly the steps used in a manufacturing scenario. The scale of civil engineering projects is so large that the construction of prototypes is not really possible. It is, however, possible to build scale models and computer simulations of the projects and to rely upon the past performance of similar projects to project a design's effectiveness.

In addition, while the manufacturing process does require conformity with certain codes, the civil engineering world is fraught with codes, standards, statutes, and regulations. If you are a civil engineer who just read the description of the hypothetical engineering design process above, you can certainly identify with the general steps. I will caution you, however, that in the first step in the process, in which you ascertain what the problem is and who needs it solved, you also need to be prepared to conduct an exhaustive review of existing codes, statutes, standards, and regulations that may influence the project's outcome, or even the project's viability.

For instance, if land use development is on the table, until the presence or absence of regulated wetlands is known, the project outcome is unknown, and perhaps its economic feasibility is at risk. If unknown hazardous waste contaminants may be present, the same uncertainty exists. In addition, statutes or regulations may be pending that, if enacted, would change the constraints on a project. These are only a few examples. At project start-up, it is imperative that all possible avenues for such constraints be explored.

Civil engineering projects range in scope from very narrow to massive. Here are just a few examples of civil engineering projects of varying scope.

Narrow scope

- Individual subsurface sewage treatment system (septic system)
- Individual residential plot plan with grading and utility connections
- Sewage or storm water pumping station

Medium scope

- Residential subdivision projects of up to several hundred homes, including planning road and lot layout, design of infrastructure (roads, storm and sanitary sewers, water distribution), lot grading, and storm water management facilities
- Small bridges and culverts servicing local roads
- Site plans for single industrial developments, shopping malls, multi-family housing

Large scope

- Sanitary sewage or water treatment plants
- Large residential developments (thousands of units)
- Capital improvement programs for municipalities

Massive scope

- Seaports and harbors
- Major tunnels and bridges (Boston's Big Dig, Golden Gate Bridge)
- Airports

The steps described for the manufacturing process are not strictly applicable to firms that participate in the narrow-scope projects listed. Many firms in that sector are simply too small to support that kind of process, because they may have only a few persons. Nevertheless, the process, to the extent practicable, should parallel the ideal. The firms at the other end of the spectrum, those that design and often participate in the construction administration phase of massive projects, may be staffed by hundreds of people, or even thousands at the multinational level. Those firms have the resources to follow stringent processes of project development and management. Modeling and testing of their projects is entirely within their capabilities. As for the firms in between, there is no exact cutoff point where the ideal engineering design process should be applied. To the extent possible, you should keep the ideal process in your mind as something desirable, then stay as close to it as you can. If you can't form a design team, you should try to achieve the design team outcomes as best you can by yourself or with a colleague.

Just as we can't always strictly follow ideal behaviors in life, so it is with the engineering design process. Where the working environment and organizational resources exist, the idealized model works exceptionally well. As you enter the engineering growth curve, do your best to follow the ideal guidelines and, when doing so simply is not possible, find ways to compensate.

24

SUSTAINABLE DEVELOPMENT

BACKGROUND

Sustainable development is a complex concept that you've likely heard about or studied, because it impacts engineering and will probably continue to do so throughout your career. Therefore, you need to understand its origins, its current status, and how it will affect the future of design and development.

Wikipedia defines *sustainable development* as a process of developing (land, cities, business, communities, etc) that "meets the needs of the present without compromising the ability of future generations to meet their own needs." One of the factors that sustainable development must overcome is environmental degradation without impeding the needs of economic development or social equity and justice. Sustainable development is said to have three "interdependent and mutually reinforcing pillars": economic development, social development, and environmental protection.

The spectrum of applications for sustainable development is incredibly wide, ranging from lofty and comprehensive plans to develop poverty-stricken third world nations at one end to sustainable development considerations in the design of a new product at the other end. What is the impact on engineering?

The American Society of Civil Engineers (ASCE) and the National Society of Professional Engineers (NSPE) have defined sustainable development as "the challenge of meeting human needs for natural resources, industrial products, energy, food, transportation, shelter, and waste management while conserving and protecting environmental quality and the natural resource base essential for future development." Infrastructure itself must be sustainable in the sense of reliably and economically meeting human needs for its vital services, in its own effects on environmental quality, and in its demands on natural resources. Additionally, the NSPE Code of Ethics in section III 2 d says:

> *Engineers shall strive to adhere to the principles of sustainable development in order to protect the environment for future generations* (www.nspe.org/ethics/eh1-code.asp).

The ASCE has issued a comprehensive Policy Statement 418, titled *The Role of the Civil Engineer in Sustainable Development*, which can be found at the end of this chapter and at *www.asce.org/pressroom/news/policy_details.cfm?hdlid=60*.

In addition, ASME and the AIChE have incorporated environmental protection principles in their codes of ethics. And it is safe to assume that more engineering organizations will do so in the future.

Historically, engineers have focused on providing the most cost-effective, efficient use of resources to fulfill the goals and objectives of the organizations they represent. Accordingly, engineers have been involved on a local rather than global basis. The notion of engineering work being a partnership of public and private interests has been the exception, not the rule. Today, the accelerating pace of technological advancement and a concern for the protection of the limited natural resources of our world mandate an increasing partnership of private and public interests. Meeting present needs without compromising the ability of future generations to meet their own needs will require that sustainable management practices—and design professionals, including engineers—must take the lead as the managers of sustainability.

In their article *Sustainable Development and LEED* (published by the Architectural Engineering Institute and the Construction Institute of ASCE) Kay Copp Brown and Lonnie Coplen describe the relationship of sustainable development to the controversial issues of global warming and the Kyoto Treaty protocol. The authors document how the regulations in a number of other countries will result in more sustainable development progress than will those in the United States. Nevertheless, the United States is making significant strides in sustainable development: U.S. colleges have introduced it into their curricula. The commercial and residential construction industries have some certification programs in place. Government and private industry are methodically working out agreements for sustainable product design and calling it product stewardship. A large number of organizations have started developing programs for voluntary or mandatory compliance programs.

IMPACT ON YOUR CAREER

At this point, you are probably thinking this all sounds wonderful and you would like to help with this concept of sustainable development. You too worry about global warming, loss of the tropical rain forest, depleting supplies of water, oil, and so on. Specifically, what can you do to support the concept of sustainable development?

On an informal basis, you can consider sustainability in your engineering designs and try to make a difference when you can. Take into consideration things like irreplaceable resources when planning projects.

At this point of sustainable design in the United States, you should research what has been done in your industry or in your area of engineering. Read the literature, attend a seminar. Become familiar with current or pending requirements of sustainable development in your work area and begin meeting those requirements now. Consider applying for certification if and when it becomes available in your area of engineering work.

IMPACT IN TWO KEY SEGMENTS

We will now review in greater detail the developments regarding sustainable design in two of the many different engineering areas.

In Building Design

Sustainable development as applied to buildings, building systems, and building materials are lumped into a category commonly referred to as Green Building Design. There is an organization that provides guidelines and success measures for sustainable engineering design in projects and products for buildings. It is the nonprofit U.S. Green Building Council (USGBC), *www.usgbc.org*. This organization conducts a project certification program called Leadership in Energy and Environmental Design (LEED). This program has become widely accepted as the standard of sustainability, or "greenness," for American buildings. While its popularity has grown at an impressive rate, there are fewer than 1,800 registered projects, representing about 5 percent of construction starts in the United States, according to the USGBC.

Developers can register their projects with the USGBC. A LEED-certified project may command a higher sale price, may sell faster, and may add credible value for environmentally sensitive investments. LEED certification requires payment of a fee and submission of considerable documentation. Furthermore, it can require significant time for document preparation. The added expenses of facilitating design documentation, energy modeling, and commissioning can add up but may not be a significant proportion of the consultant fee. These added expenses can sometimes be at least partially offset by grant or incentive programs. For larger companies, the LEED certification costs can be readily absorbed. For smaller engineering firms, this certification can be cost prohibitive.

An alternative approach to green design involves developing a project that meets LEED performance goals but stops short of filing for LEED project certification unless the owner is sold on the value of the documentation and explicitly requires it.

LEED certification was initially developed for office building applications. In 2004, it was expanded to include existing buildings and construction/renovation of tenant space. USGBC is also adapting LEED to different market segments, such as health care, schools, and laboratories, by issuing application guides that focus on addressing specific challenges uniquely relevant to these types of projects.

You and your company should be aware that there are serious professional liability concerns regarding sustainable development green building standards:

- Green buildings often use new materials that have not been in long-term use. Many liability laws have a longer statute of limitations for the design professional than for the manufacturer. You and your firm need to investigate this issue in the states where you design all or part of a green building.
- Green buildings use innovative concepts that may result in what has been called the "sick building syndrome, " which can also place the design professional at risk for possible malpractice and other claims.
- Green buildings need an independent expert to ensure that the building is properly commissioned, operated, and maintained. The building management's staff must be properly trained. If the building is not operated and maintained properly, the design professional may be at risk for liability claims.

The USGBC also has a certification process for design professionals that leads to the LEED Accredited Professional Certificate. The purpose of the exam and certificate is to ensure that a successful candidate has the knowledge and skills necessary to participate in the design process, to support and encourage integrated design, and to streamline the application and certification process. It is also to test understanding of green building practices and principles and familiarity with LEED requirements, resources, and processes.

The International Institute for Sustainable Design, *www.iisd.org*, deals more with the nonengineering aspects of sustainable development.

In Manufacturing and Product Design

Within the manufacturing and product design economic sector, there are two forms of application of sustainable development: industrial ecology and product stewardship.

Industrial ecology refers to the exchange of materials between different industrial sectors where the waste output of one industry becomes the feedstock of another. For example, the excess steam from an electrical generating facility can be used as a heat source for a nearby chemical manufacturer. The fly ash from a coal-fired generating station can be used as an input for the concrete industry.

Industrial ecosystems refer to situations in which a number of different companies, usually in close proximity to each other, exchange a variety of waste outputs. This is a relatively new and leading edge paradigm for business. It emphasizes the establishment of public policies, technologies, and managerial systems that facilitate and promote production in a more cooperative manner. Implementing industrial ecology involves such things as life cycle analysis, closed loop processing, reusing and recycling, design for environment, and waste exchange. Technologies and processes that maximize economic and environmental efficiency are referred to as ecoefficiency.

Product stewardship is a term used to describe a product-centered approach to environmental protection. It calls on those in the product life cycle, including designers, manufacturers, retailers, consumers, waste managers, and disposers, to share responsibility for reducing the environmental impact of products.

Product stewardship activities have been taking place globally for over a decade. In the United States, this idea is now gaining interest as more state and local governments cope with large, ever-changing, and complex waste streams.

Besides dealing with waste management issues, product stewardship is about reducing the impact of products as far up the product chain as possible. This can result in conservation of resources and protection of air, land, and water. The goal is that product stewardship drives better design and resource efficiency, which saves money for the producer.

Many product stewardship efforts are now focused on end-of-product-life management issues. Long-term goals are to affect the design stage of products so that they are less toxic and more readily refurbished or recycled. For example, agreements have been reached between government and the electronics industry and with the carpet industry to collaborate in finding ways to manage their wastes, capture and reuse the materials, and improve future product design. In response to product stewardship requirements in Europe and Asia, the manufacturers of computers are now making products that are easier to recycle by limiting the types of plastics used, labeling computers, and designing them to be easily taken apart and upgraded.

Historically, government and the taxpayers have borne most of the cost and responsibility for the management of products at the end-of-life. The concept of product stewardship seeks to put more of that cost and responsibility onto those who produce and consume those products.

Product stewardship can lead to improvements in air and water quality as well as in the waste management system. It is a new approach to solving existing problems. True product stewardship is an extension of pollution prevention, technical assistance, and green purchasing efforts. It makes economic and environmental sense and can help move us toward sustainable communities and resources.

The Product Stewardship Institute (PSI) is a national nonprofit membership-based organization. PSI works with state and local government agencies to partner with manufacturers, retailers, environmental groups, federal agencies, and other key stakeholders to reduce the health and environmental impact of consumer products. PSI takes a unique product stewardship approach to solving waste management problems by encouraging product design changes and mediating stakeholder dialogues.

THE ROLE OF THE CIVIL ENGINEER IN SUSTAINABLE DEVELOPMENT
ASCE Policy Statement 418

Approved by the Committee on Sustainability on June 17, 2004
Approved by the National Infrastructure and Research Policy Committee on July 20, 2004
Approved by the Policy Review Committee on July 23, 2004
Adopted by the Board of Direction on October 19, 2004

Policy

Sustainable development is the challenge of meeting human needs for natural resources, industrial products, energy, food, transportation, shelter and effective waste management while conserving and protecting environmental quality and the natural resource base essential for future development.

The American Society of Civil Engineers (ASCE) recognizes the leadership role of engineers in sustainable development, and their responsibility to provide quality and innovation in addressing the challenges of sustainability. The ASCE Code of Ethics requires civil engineers to strive to comply with the principles of sustainable development in the performance of their professional duties. ASCE will work on a global scale to promote public recognition and understanding of the needs and opportunities for sustainable development.

Issue

The demand on natural resources is fast exceeding supply in the developed and developing world. Environmental, economic, social, and technological development must be seen as interdependent and complementary concepts, where economic competitiveness and ecological sustainability are complementary aspects of the common goal of improving the quality of life.

Sustainable development requires strengthening and broadening the education of engineers and finding innovative ways to achieve needed development while conserving and preserving natural resources.

The achieve these objectives, ASCE supports the following implementation strategies:

- Promote broad understanding of political, economic, social and technical issues and processes as related to sustainable development
- Advance the skills, knowledge and information to facilitate a sustainable future, including habitats, natural systems, system flows, and the effects of all phases of the life cycle of projects on the ecosystem
- Advocate economic approaches that recognize natural resources and our environment as capital assets
- Promote multidisciplinary, whole system, integrated and multiobjective goals in all phases of project planning, design, construction, operations, and decommissioning
- Consider reduction of vulnerability to natural, accidental, and willful hazards to be part of sustainable development
- Promote performance-based standards and guidelines as bases for voluntary actions and for regulations in sustainable development for new and existing infrastructure

Rationale

Engineers have a leading role in planning, designing, building and ensuring a sustainable future. Engineers provide the bridge between science and society. In this role, engineers must actively promote and participate in multidisciplinary teams with other professionals, such as ecologists, economists, and sociologists, to effectively address the issues and challenges of sustainable development.

25

PROJECT DELIVERY SYSTEMS

This chapter will introduce you to a variety of project delivery systems. A *project delivery system* is the organizational structure of the various participants who will fulfill certain roles on a project and have certain obligations toward bringing an owner's goals and objectives to a fruitful conclusion. Before we discuss various project delivery systems, let's give some definition to the various participants who make up a team.

- *Owner.* The person, persons, or entity that sets the goals and objectives for a construction project and who will supply the financial resources to accomplish it
- *Contractor.* The person, persons, or entity obligated to furnish labor, equipment, materials, and all else necessary to construct a project in accordance with the contract documents (plans, specifications, construction contract, and applicable law)
- *Design professional.* A person or entity legally authorized to practice as an architect, engineer, land surveyor, or landscape architect in the state where a project is to be constructed. In the context of this chapter, the design professional will likely be an engineer or engineering company. The scope of service furnished by the design professional may include predesign studies, preparation of conceptual or planning documents and/or studies, development of plans and specifi-

cations for the construction of a project, preparation of opinions of probable construction cost, bid documents, contract documents, and so forth. The design professional generally provides assistance to the owner in applying for environmental permits and land development approvals from local, county, and state agencies of jurisdiction, and will interact with various utility companies on behalf of the owner. The design professional will also furnish construction-related services, such as construction contract administration and observation or inspection of construction.

- *Construction manager.* A person, persons, or entity that provides various forms of guidance or assistance to an owner. The scope of service of a construction manager will be contractually determined. Services may be furnished during the design process before the onset of construction, but generally the construction manager will be engaged to manage the actual construction process. A construction manager may be a design professional but is usually someone with experience as a construction contractor.

Project delivery systems are applicable to construction projects of various sizes, although for small projects in the private sector, the project delivery system may be more relaxed than those we discuss in this chapter. Some clients in the private sector may completely shy away from written contracts with the design professional, as well as the preparation of comprehensive project specification and construction contracts used with the eventual contractor. It has been my experience that such projects can be brought to a successful conclusion, but that they can also produce a myriad of misunderstandings and claims of professional negligence. To the extent possible, design professionals should strive to overcome the owner's reluctance to formalize the process. When that effort fails, the design professional has to evaluate the risk versus reward potential, make a well-reasoned business decision, and either accept or reject the project.

This chapter will likely be of most interest to engineers in private practice acting as design professionals, to engineers in government and industry who are acting as owners, and to engineers in construction acting as contractors.

There are several prominent categories of project delivery systems, some with subcategories or variations on the main theme. We will explore those that are used predominantly and mention briefly the subcategories. The various prominent categories are these:

- design-award-build (DAB), also known as design-bid-build (DBB)
- design-build (DB)
- turnkey (a variation of design-build)

The choice of project delivery system must be made by the owner. The owner may want to consult with the design professional in an effort to become more familiar with the advantages and disadvantages of each type of project delivery system, but in the final analysis the owner must decide.

In the case of an owner that is an industrial manufacturing entity, the owner may want its own employed engineers to participate with the consultant design professional. For instance, they may be asked to design machinery, designate its placement within a facility, design electrical and mechanical control systems peculiar to the industry, and so on. A detailed discussion of that special circumstance is beyond the scope of this book, but you should be aware that it might happen from time to time. Special attention must be paid to the contract form for the design professional to have appropriate protection from any liability that may arise because of the negligence of the owner's staff. Because the seal of a design professional on a set of drawings means that he or she has supervised the preparation of the plans and takes personal responsibility for the quality and completeness thereof, competent and experienced counsel and the insurance carrier should carefully review this situation.

We will look at some of the advantages and disadvantages of the various project delivery systems in this chapter, but this discussion will not be exhaustive. You must make an effort to gain more knowledge through independent research, by closely following developments in the construction world, and by monitoring the expanding body of knowledge. One of the areas from which changes arise is common law, the findings of higher courts (appeals courts and supreme courts) in the various states and at the federal level.

A reminder: You can learn about legal findings by reading them, but stoutly refuse to give legal advice. That's reserved for attorneys.

DESIGN-AWARD-BUILD (DAB)

DAB has been by far the predominant project delivery system in use in the United States, although it is gradually giving way to design-build (DB). It is important to recognize that there are two types of owners: private sector and public sector. Public sector owners are much more constrained by statutes and regulations from state to state than are those in the private sector. Both types of project delivery systems (DAB and DB) for public owners will be linked to laws in their states. Some states require DAB and make no provisions for DB, while other states may permit but not require DB. In the private sector, owners are free to select the project delivery system of choice. DB has also been growing in use by federal agencies on their construction projects.

In a typical DAB project, the owner will develop the goals and objectives of the project as well as a proposed budget. For instance, an owner (public or private) may envision construction of a small dam across a river tributary to create a recreational lake above the dam. The concept may call for parking lots, boat launch areas, lighting, water distribution, comfort stations, and a playground area. There might be a $2.5 million budget established for the project. That concept may become articulated in a request for proposal (RFP).

The owner will use the RFP to inform prospective design professionals of the project's goals and objectives. Hopefully, the owner will attempt to procure the design professional using qualification-based selection procedures, as described in Chapter 32. After the proposals have been returned by the design professionals, the owner will choose a firm for the job. Once the design professional is under contract, the engineering design process (described in Chapter 23) will commence. The design professional will work with the owner through the planning and design process, eventually delivering a set of plans, specifications, and contract documents to the owner. As soon as all necessary permits and approvals from outside agencies have been attained, the project will be ready for the bidding phase.

During the course of the design professional's work, the professional may have regular contact with the owner, who will follow the development of the project plans and give input to the project team. Perhaps the owner will attend regularly scheduled project meetings. When all parties are satisfied with the design, presuming that the design professional's opinion of probable construction cost is within budget, the project will be ready for bidding. If the budgetary constraints have not been met, some adjustment in the design or in the budget, or both, may be required, and then the project will be ready for bidding.

The design professional will assist in the solicitation of bids. If the owner is a public agency, the method of solicitation will be prescribed by law. For private sector work, the design professional may contact construction contractors of known reputation for reliability and skilled work, and ads may be placed in newspapers with local and regional coverage. The design professional will assist the owner in receiving bids and checking them for compliance with the bidding specifications. Usually the lowest responsible bidder will prevail, although sometimes exceptions occur that are beyond the scope of this book.

When the construction contractor has been selected, the owner will enter into contract with that entity. At that point, there will be contractual relationships between the owner and design professional and between the owner and construction contractor but not between the design professional and the construction contractor. The owner will be the superior party in both contracts and will exercise the most control. The design professional may have some contractual duties during construction, and the role of the design professional during construction may be articulated in the owner-contractor document, but the design professional has no direct contractual duty to the contractor, or vice versa.

The design professional's duties may include construction administration on behalf of the owner, periodic observation or even resident inspection duties, running periodic construction project meetings, certifying contractor's periodic vouchers for partial payment, processing shop drawings, answering requests for information, interpreting the plans and specifications whenever necessary, and monitoring the contractor's work for compliance with the contract documents. The design professional is never responsible for the contractor's ways, means, and methods of prosecuting the work and must astutely avoid taking on even the slightest duty with respect to that. Eventually the project will be completed and accepted by the owner, and the parties will fold their tents and leave.

The DAB project delivery system proceeds in a step-by-step manner, often referred to as linear. The roles of the parties should be clear. The responsibilities of all parties should have been set out in the various contracts and in the contract documents. The advantages of the DAB process include the following:

- The owner has control.
- The owner has an opportunity to participate and contribute ideas as the design process unfolds and can object, in a timely fashion, to design features that are undesirable.
- The owner can contractually delegate authority to the design professional to act as the owner's agent during construction.

- The owner will get the best price through the bidding process.
- The roles of the parties are clearly defined.

Some of the disadvantages of DAB are these:

- The construction contractor has no input during the design phase.
- DAB projects are difficult to fast track. Fast-tracking would require construction to get underway before the design is completed.
- Too often, DAB sets up an adversarial relationship among the parties if difficulties arise during construction (errors and omissions, differing site conditions, lack of availability of specified products, etc.).
- Bids may come in above the design professional's opinion of probable construction costs, putting the project above budget.
- Project time may be extended because materials or equipment that require long lead times could not be ordered in advance of awarding the construction contract.

DESIGN-BUILD

The DB project delivery system has been used for centuries, even millennia. The ancient Egyptians created the pyramids under that system. The Romans and Greeks built magnificent edifices that way many centuries ago, many of which stand today to remind us what good engineers, architects, and builders they were.

DB is a project delivery system by which the functions of the construction contractor and the design professional are combined into a single unit, the design-builder. From the owner's perspective, DB is much simpler. The owner will only need to deal with a single party. That party, the design-builder, will be responsible for performing all of the design and construction necessary to fulfill the owner's project scheme.

The owner will enter into contract with the lead entity of the DB team, with other members being subcontractors to the lead entity. DB teams can be formed in a variety of ways. The lead partner of the DB team can be the design professional, with the contractor engaged as a subcontractor to the design professional. In the most prevalent structure, the roles are reversed, with the contractor as lead entity. Only a very large design professional firm with a lot of financial resources can be the lead. In a third format, the contractor and design professional may form a joint venture, a company formed expressly for the project at hand. Then both the contractor and design professional will become subcontractors to the joint venture. Whatever the form, the owner will have a contractual relationship only with the design-builder but not with the partners of the design-builder team.

Considerable concern has arisen in the professional engineering community about the role of the design professional in the DB process. Professional engineers, by law and by ethical codes of conduct, have a paramount duty to protect the health, welfare, and safety of the public. There has been concern that the professional engineer who is a member of a DB team may have pressure exerted by other members of the DB team to cut costs by skimping in design, or that the professional engineer's advice may be overruled. Another concern is that the professional engineer has a duty to the client, ordinarily the owner, but in the DB project delivery system, the client is the design-builder. The National Society of Professional Engineers (NSPE) adopted a position statement in 1995 relevant to this concern, "Position Statement No. 1726 – Design Build in the Public Sector." While neither supporting nor opposing the DB concept, NSPE does suggest that the owner/design-builder contract should contain language requiring that "in all matters that affect the health, safety, and welfare of the public, the designer must be the decision maker," and that "engineering services are required to be performed by or under the supervision of a licensed professional engineer."

Now let's look at how a typical DB team will function. Generally speaking, the owner is seeking a single source for all services related to making the project happen through design and construction. That relieves the owner of the need to separately engage a design professional and a contractor, thereby relieving the owner of the need to coordinate the design and construction services. It also gives the owner a high degree of comfort, knowing that contractor/design professional disputes that may arise will be resolved within the design-build entity without owner involvement. Additionally, in virtually all design-build contracts, the owner will delegate many of the responsibilities and much of the authority normally retained in the DAB project delivery system.

With respect to DB contracts, the Engineers Joint Contract Documents Committee has created a family of contract documents exclusively for the design-build project delivery system. You can check out these contracts at *www.nspe.org/ejcdc/home.asp.* Model DB contract documents are available at the following Web sites:

- American Institute of Architects (*www.aia.org/docs_newtitles*)
- Design Build Institute of America (*www.dbia.org/pubs/contracts.html*)

It would be preferable for engineering firms to use the EJCDC contract documents, but the other forms may be proposed by the owner or by members of the DB team.

In the DB project delivery system format, the engineer and the contractor can work together during design. As a result, such things as choice of materials and equipment will be largely influenced by the contractor. A big benefit to the engineer and to the project will be the contractor's input as to constructability, something usually not available to engineers in the DAB format. The owner will have little or no control over the design, however, presenting the risk that the owner will be disappointed if the final product does not fully meet its expectations. That would probably mean the owner's program documents at the beginning of the project were not sufficiently complete to ensure a satisfactory outcome.

A modification to the DB team structure defined above is termed "bridging." When bridging is used, it means that the owner will engage a design professional who is not part, of or related to, the DB team to serve as the owner's consultant. The independent design professional will help the owner with program development and preparation of a request for proposal (RFP) that will clearly and thoroughly define the project goals and objectives. He or she will be available to monitor and review design documents being prepared by the DB team and may give input on behalf of the owner to the DB team's design professional. The design professional may also review payment requisitions submitted by the design-builder during the course of the work. This type of structure significantly enhances the owner's level of confidence in the project and provides invaluable oversight by a design professional whose only allegiance is to the owner.

The DB project delivery system has advantages and disadvantages. Advantages include the following:

- The owner has only a single entity pursuing the successful design and construction of the project. It does not matter which partner bears responsibility for an error or omission. The DB team members will have to resolve that among themselves.
- Design and construction will be completed in a shorter time than under the DAB project delivery system.

- The cooperative environment in which the design professional and contractor work should benefit the project. After all, they chose each other at the outset, so they must feel some form of business respect for each other.

Disadvantages include these:

- The owner loses control over the project during the design phase.
- Design professional representation is absent for the primary interest of the owner, though this can be relieved by bridging.
- The outcome may not meet the owner's expectations.

TURNKEY PROJECT DELIVERY SYSTEM

A turnkey project delivery system is very similar to a DB process. It goes beyond the construction of a facility, at which time the DB team can break up and go their separate ways. In a turnkey project delivery system, the design-builder will be engaged not only to design and construct the project but also to start it up and to operate it for a specified period of time. When that time expires, the design-builder will turn the keys over to the owner (therefore, "turnkey"). Turnkey project delivery systems take various forms:

- design-build-operate-maintain
- design-build-operate-transfer

In some additional forms of turnkey project, the design-builder may be asked to finance the project as well.

CONSTRUCTION MANAGER

The owner may further expand the level of service it will receive under the DAB process by engaging a construction manager (CM), usually a general contractor. The CM can be engaged at the front end during the design phase and will represent the owner's interest by developing construction cost estimates as the project proceeds, by doing constructability reviews at appropriate stages of completion, by analyzing various project components for value, and by furnishing invaluable advice to the designer from a contractor's point of view. During this process, however, the design professional must realize that he or she is responsible for the project design, the CM's advice notwithstanding; if something is on the plans or in the specification prepared by the design professional, he or she will bear personal responsibility for the work.

At the commencement of construction, the CM may act as general contractor and in that role be responsible for hiring, supervising, coordinating, and controlling the subcontractors. CMs who do this kind of work are generally known as "CMs at risk," meaning that they are financially liable for any losses the owner may incur as a result of the work of the CM and the subcontractors.

Another form of CM is the agency CM. An agency CM will be engaged by the owner to provide advice during the project. An agency CM will not manage the construction project in the same manner as the CM at risk but will be available to represent the owner's interest during the project. An agency CM may be a contractor or a design professional.

As you can see, there are many types and forms of project delivery systems. No matter which type of project delivery system you may experience during your career, always remember that you are an engineer, a member of a noble profession dedicated to improving the lot of humanity. You are duty-bound to keep the health, safety, and welfare of the public paramount in your practice. Review the ethics section of this book from time to time, and, by all means, read the Engineer's Creed often. Some engineers actually carry business-card-sized copies of it in their wallets.

26

CONSTRUCTION PROJECT PROCEDURES AND TEAM ROLES

In this chapter, we paint a fundamental picture of the way a construction project is conceived, planned, and executed and give you insight into the role that you and your employer will play. This example is largely related to engineers who will work in the consulting industry, as well as those who may become employed by owners, such as government agencies, land developers, or industrial organizations that may ask you to be their representative to a construction project.

The procedural path for an engineering project will vary from firm to firm and from project to project. In Chapter 25 on project delivery systems, we described multiple variations of types of systems. In this chapter, we will focus on the most common form of project delivery system, design-award-build (DAB, also known as design-bid-build, or DBB).

In a design-award-bid scenario, the key parties are the owner (client), the design engineer, and any subconsultants or experts engaged by the engineer in support of the project. The role of the owner is to set the goals and objectives of the project and to finance its design and execution (construction). The engineer's role is to furnish predesign services (studies, project-planning documents, conceptual alternative documents, preliminary opinions of probable construction costs), plans and specifications for construction of the project, and contract documents (contract, general conditions, etc.).

A PROJECT BEGINS

A typical project begins with an owner's perceiving a need and seeking an engineer's assistance to develop the details and construction plans to convert the concept to reality. The owner will prepare a request for proposal (RFP) and send it to various engineering firms. The RFP may also be offered through advertising in local or regional newspapers. In some circumstances, the owner may solicit bids from engineering firms for the work. The preferable practice is for the owner to solicit qualification of the firms and rank them based upon a screening of the qualification information. Then the owner can call in the top-rated firm and negotiate toward a mutually agreeable fee. Failing that, the owner can move on to the next highest-ranked firm, and so on, until an agreement is reached. That process is known as quality-based selection (QBS), which is discussed in more detail in Chapter 32.

After the engineer is engaged, through execution of a mutually agreeable contract for professional services, the engineer will begin detailed project planning. In the planning stage, the engineer will assemble a project team within the firm, staffing it with people who have the required skills and experience relevant to the project. It may be necessary to engage outside experts or subconsultants to staff the project team adequately.

It is extremely important that the owner's objectives, timing, and budgetary constraints be clearly identified at the outset, along with any other criteria that are critical or mandatory in the owner's planning.

DEVELOPING AND EVALUATING ALTERNATIVES

As the project gets under way, the engineer's contractual duty may be to develop and evaluate conceptual alternatives for achieving the project's goals and objectives. The alternatives will be developed considering a multitude of impacts, such as environmental laws and regulations, relationship of existing facilities, public safety, operational considerations (staffing, costs), and more. The conceptual alternatives are not prepared in a detailed manner. They must all, however, meet the project constraints set by the owner.

When the project has been successfully considered for conceptual alternatives, the owner, in conjunction with advice from the engineer, will select one for implementation. The project team will meet with the designated project manager who will serve as team leader. The overall project, including the owner's goals and objectives and the manner in which the team will approach the project, will be discussed. Typically, members of the design team will be given specific tasks along with time lines for completion and will be asked to commit to the schedule. At that point, everyone must be frank and candid. The project manager will not know of the team members' individual workloads and will rely on them to resolve individual conflicts of schedule.

THE ACTION PLAN

An action plan document should be prepared in which the project activities are listed along with the names of the responsible parties and the expected time of completion. As the project proceeds, the initial timelines for some activities may be disrupted because of unforeseen occurrences, failure of the owner or subconsultants to deliver necessary documents or services, or other events. Periodic meetings of the project team will reveal these as they arise, allowing the project manager to make adjustments to the action plan to keep the project on track and within budget.

If you are on a project team that has fallen behind or has a budgetary problem, you may be asked to work overtime, especially if you are an "exempt employee." (An exempt employee is a professional on salary who receives no compensation for overtime. The actual definition is from the U.S. Fair Labor Standards Act and subsequent rulings.) We advise that you do so cheerfully, without grousing. The firm has a lot at stake when project schedules or budgets are broken.

It is essential that, during the design process, effective project controls be in place and functioning. Quality assurance/quality control (QA/QC) is an important responsibility of the project manager. This can mean pairing the right personnel with specified tasks, managing the number of personnel assigned to each task, knowing (or learning) the specific skills and experience of individual team members, and ensuring good coordination between team members whose tasks are interrelated.

If all goes well, the design phase will finish on time and within budget. The term for the final phase of design is "close-out." The work of the design team members will be collected and assimilated by the project manager, who will then have the responsibility of making a final completeness review. If it is considered complete, it will be turned over to the owner for bidding or for negotiations with a selected contractor.

THE DESIGN TEAM

Let's talk about the design team itself. It will be led by a project manager, usually a well-experienced engineer, who is fully aware of the owner's goals and objectives. The project manager will be responsible to upper management for the preparation of a budget to perform the design work and for maintaining the budget. The project manager will also be responsible for making and adhering to the project schedule; for selecting team members with the requisite skills, knowledge and experience; and for maintaining project quality. The project manager will meet periodically with the owner to review the project as it progresses, to manage contract amendments resulting from scope changes by the owner, and to practice risk management for the firm.

The members of the project team may include engineers, draftsmen, surveyors, structural designers, environmentalists, and administrative support. The best project teams are composed of personnel who know what *team* means: "It's not about me; it's about us meeting our goals." A good practice is for the project manager to hold periodic project team meetings. This is particularly true if the project requires multiple disciplines or specialties, such as site work, structural, electrical, hydrologic, soils, and the like. Coordinated design functions and sharing of information are critical to all projects and are even more important if various disciplines are working in their own separate work environment.

As indicated earlier, the owner commonly engages the design engineer to provide professional services during construction of the project. The role of the design engineer will be determined by contract. No single model defines the design engineer's role. We will look at a couple of possibilities.

Before the project actually goes to construction, a signed contract between the owner and a construction contractor must exist. Most often, that contract is drafted collaboratively between the owner and the design engineer, and this proposed construction contract will become part of the RFP. Contractors wishing to respond to the RFP and submit bids for the project will be required to obtain a complete set of contract documents, including plans, technical specifications, general conditions of the contract, and the proposed construction contract that the successful bidding construction contractor will be expected to sign. The construction contractor is obligated to make a thorough review of all the documents and base the bid upon the understanding obtained from that review.

When all of the documents necessary for bidding are prepared for distribution to bidders, the owner, with the design engineer's assistance, may advertise for bidders. If the owner is a public entity, the advertising procedure will always be used. Private owners, however, have the option to distribute the documents to known contractors. Sometimes the design engineer will assist the owner in finding contractors of known reputation possessing the required experience.

The design engineer will assist the owner by receiving bids and opening them at a specified time, reviewing submitted bids for conformity with the bidding requirements, checking for any informalities or deviations from the bidding requirements, checking for the presence of properly prepared surety bonds that may have been required of the contractor, checking arithmetic calculations in the case of unit price bids, and so on. When that process is complete, the design engineer will report to the owner the results of the bidding process, usually recommending the award go to the lowest responsible bidder. The owner will execute the contract (it should already have been executed by the contractor with the bid submission). The project will then transition to the construction phase.

Depending on the nature of the design engineer's contractual duties and the size and complexity of the project, the design engineer will assemble an in-house construction team. A larger construction team will consist of the various parties to the contract, including the owner, the contractor and subcontractors, and the owner's resident project representative (RPR) whose role we'll discuss later. Some design engineering firms have formal construction departments that focus all of their attention on construction operations. Assuming that is the case for our hypothetical project, personnel particularly suited to the project at hand will be selected for the team.

Typically, a licensed professional engineer with significant construction experience will head up the design engineer's construction department. In that case, the department head will designate a project engineer to head up the team for the project, and they likely will work together to select technicians to work in the field.

Many engineering firms have personnel who have been trained and certified by the National Institute for Certification in Engineering Technologies (NICET). NICET is an affiliate organization of the NSPE. The NICET Web site is at *www.nicet.org*. NICET certification is widely recognized as solid evidence of good credentials in the field. In addition to the specialists from the construction department, a member or two from the original design team may be on the team to act in a consulting or advisory role, as the field construction people are seldom, if ever, involved with the design phase of the project and will need the insights of the design team member as the project proceeds.

For projects of considerable complexity and/or size, the owner may choose to delegate authority to a construction manager, who may be a member of the owner's staff or a person or firm that specializes in construction management. That entity will serve as the owner's RPR (resident project representative). In that scenario, the design engineer's construction team will act as an advisory resource for the RPR. In some circumstances, however, the owner may choose to delegate authority to the design engineer to serve as owner's RPR, a significantly more demanding role.

Let's look at some common duties of the design engineer's construction team when the owner designates a third party as RPR. It is important to know that when the owner delegates its authority to an RPR, the design engineer then acts only as an advisor to the RPR and has no authority over the contractor. These are some of the traditional services provided by the design engineer:

- Providing clarification of contract documents whenever questions arise as to intent, or if there appears to be a conflict in the documents
- Reviewing shop drawing submittals for conformity with the contract documents and requesting revisions or corrections when necessary
- Reviewing change order requests, along with any associated costs
- Acting as consultant to the RPR with respect to technical elements of the design
- Acting as consultants to the RPR on progress of the project and on the quality of construction
- Reviewing and approving or rejecting progress payment requests by the contractor

- Observing the quality of the construction progress on behalf of the RPR
- Providing interface with regulatory agencies

If the design engineer is designated as the owner's RPR, then the owner's authority will be passed to the firm. In that circumstance, the design engineer will be responsible for all the duties listed above but will no longer be serving as an advisor or consultant to the RPR. The design engineer will have substantially more authority as well as responsibility under that format. We believe that none but the brave hearted and highly qualified engineering firms should undertake such a task.

This might be a good place to point out that the design engineer should never take on the responsibility for construction site safety, except as required by law for its own personnel. The body of law dealing with construction site injuries and death is changing as time goes by, and it is changing in a state-by-state manner. While it is beyond the scope of this book, we urge you to be sensitive to the issue of construction site safety and to be absolutely sure you know what your firm's policy is with regard to that matter. Safety of construction workers and of third parties at a construction site is no laughing matter, but it should be the responsibility of the owner and the contractor to maintain safe site conditions and to comply with laws regarding worker safety for their employees.

During the course of the project, project meetings will be held at the construction site, usually on a regular basis. If you are in attendance at such meetings, pay a lot of attention to what is said. Keep personal notes, and if you are given the opportunity to review the meeting minutes, report anything that you think is erroneous to your construction team leader. It is also important to report anything you remember that was omitted from the minutes, particularly if you have notes regarding the omitted information. Your supervisor will decide what to do from that point.

Under normal circumstances, the design engineer will have no authority or responsibility over the contractor's ways, means, and methods of construction. It is very bad practice for a member of the construction team to correct the contractor's personnel in their work. If there is a deviation from good practice, or if the project plans and specifications are being violated, that should be reported to the owner's RPR. If that happens to be the design engineer, then the reports should be made to the leader of the engineer's construction team.

BRINGING THE PROJECT TO FULFILLMENT

At some point, the project will be substantially completed. Punch lists (lists of things that need to be corrected by the contractor) will be prepared, and the project should be made ready for final inspection. The contractor will be required to deliver any warranties or guarantees on equipment or materials furnished, operation manuals, as-built drawings, and maintenance bonds.

If the project has been well managed by the owner's RPR, the contractors, and the design engineer's construction team, everyone can walk away with a feeling of satisfaction and achievement. In most projects, stumbling blocks come up along the way. If everyone is striving for a quality finished project brought in on time and on budget, those stumbling blocks can usually be handled expeditiously. Projects tend to get a little unwieldy when the parties become more interested in their own outcomes that the outcomes of the project. It will take some time for you, should you become engaged in construction teams, to master the most appropriate behavior. Maintain your integrity, and don't be afraid to speak up. Preferably, do your speaking within the framework of your firm's construction team, leaving the resolution of the bigger problems to the team leader. Pay attention to the manner in which scenarios play out, because someday you may be asked to lead the team.

In Chapter 9, we discussed leadership and management. The distinction between the two is that managers concentrate on getting things done, while leaders are goal setters, inspirational figures who can motivate people to accept and strive for the goals. In the context of the project team, everyone can practice a degree of leadership. The project manager, however, carries the heaviest load in the project and must focus on getting things done on time, on budget, and within appropriate levels of quality. If the project manager is also a good leader, he or she will vastly enhance the probability of project success.

If you aspire to become a project manager or department head, we advise you to focus on developing leadership skills through training, reading, and practicing leadership in various situations. One good place to do that, as we have said elsewhere, is in a professional or technical society. The society environment offers young engineers many opportunities for leadership training.

27

PROJECT
MANAGEMENT

In the beginning of this book, we said that much of engineering is performed in projects. In Chapter 23 about the engineering design process, we walked you through a simple example of an engineering project. In Chapter 25 about project delivery systems, we reviewed the major types of civil or construction project structures—namely design-award-build, design-build, and turnkey. In Chapter 26 about project procedures and team roles, we discussed procedures for executing an engineering project in a consulting engineering firm. These procedures are distinctive and include the following steps:

1. Define the owner's project needs
2. Develop a basic plan that is used for an RFP (request for proposal)
3. Select a professional service provider
4. Define in detail the project's objectives, timing, and budgetary constraints
5. Review possible conceptual alternatives
6. Designate a project manager and project team members
7. Develop a detailed project action plan
8. Implement a system of project controls

9. Implement the project

10. Complete the project on time and within budget to the owner's satisfaction

Within each of these steps, there are many skills to be learned, theories to be considered, and tools that may be used. Project management is a very complex subject about which many books have been written. Some engineers and even engineering firms provide only project management services. There is even a Project Management Institute (PMI) that defines itself as follows:

Vital and forward thinking—focused on the needs of project management professionals worldwide; that's the Project Management Institute of today. We've long been acknowledged as a pioneer in the field and now our membership represents a truly global community with more than 200,000 professionals, representing 125 countries. PMI professionals come from virtually every major industry, including aerospace, automotive, business management, construction, engineering, financial services, information technology, pharmaceuticals, health care, and telecommunications. (*www.pmi.org*)

Project management skills, while very important to the engineering process, are also used extensively beyond engineering applications.

If you are just starting your career, it is unlikely that you will immediately be assigned as a project manager. Then why review project management concepts now? You may not be placed in charge of a project to design a new hybrid automobile, but you may be assigned to develop a component as a part of a project team. You may not be placed in charge of constructing a new senior citizen housing project, but you may be assigned to develop a site drainage scheme. Your job may only be a small piece of the pie, but it should be considered to be a miniproject. Manage your small piece well, and it will lead to more challenges and growth opportunities. As your engineering career advances, you will need to know more about project management because you will become a part of a project team. You may even be asked to manage a project team.

For these reasons, we will review project management in a bit more depth, recognizing that a complete discussion of project management is well beyond the scope of this book.

DEFINITION OF PROJECT MANAGEMENT

As you may imagine, multiple definitions exist. The definition provided by PMI is:

The application of knowledge, skills, tools, and techniques to project activities in order to meet or exceed stakeholder needs and expectations from a project. Meeting or exceeding stakeholder needs and expectations invariably involves balancing competitive demands among:

- Scope, time, cost, and quality
- Stakeholders with differing needs and expectations
- Identified requirements (needs) and unidentified requirements (expectations)

The depth of the subject of project management is described by PMI as follows:

Project management is the application of knowledge, skills, tools, and techniques to a broad range of activities in order to meet the requirements of a particular project. Project management is comprised of five Project Management Process Groups—Initiating Processes, Planning Processes, Executing Processes, Monitoring and Controlling Processes, and Closing Processes—as well as nine Knowledge Areas. These nine Knowledge Areas center on management expertise in Project Integration Management, Project Scope Management, Project Time Management, Project Cost Management, Project Quality Management, Project Human Resources Management, Project Communications Management, Project Risk Management, and Project Procurement Management.

You would be ill-advised to try to master all these aspects of project management early in your engineering career unless you are specifically assigned to do so. But you should be aware of the scope of the subject and know how to investigate further should a need arise.

THE NEED FOR PROJECT PLANNING AND CONTROL

One of the general characteristics of projects is that project size and complexity seem to grow over time. With the more rapid pace of development activities caused by today's international competitive environment, timely completion has become even more important. Today we are faced with increasing pressures to complete projects faster and with lower costs.

The value of the project plan lies in its implementation. When implemented effectively, progress is measured against the planned targets, and any schedule or cost deviations are corrected. If the project plan cannot work, then the plan is modified. The project is kept on schedule and within budget through the control function, a key part of the project plan.

The effectiveness by which an organization manages its projects is the key to developing and retaining clients or customers, which in turn is crucial for the ongoing profitability of the organization. Your successful contribution to the project will provide you with much satisfaction with the project team's and the organization's accomplishment.

PROJECT TASK MANAGEMENT— A BRIEF HISTORY OF TECHNIQUES

Activity List

The most simple project management tool is a list of activities in chronological sequence with a time estimate for each activity. This tool can even work where a few activities happen simultaneously. Table 27.1 shows some of the major steps in building a house.

It would be difficult to use this tool if we are building not one house but a 30-house development. Then we would have to add all the site development activities as well as multiple simultaneous activities.

TABLE 27.1   *Sample Chronological Sequence of a Project*

Activity	Estimated Time in Weeks
1. Excavate lot	0.5
2. Form and pour basement slab and walls	1.0
3. Order long-lead-time items	0
4. Frame walls and roof	2.0
5. Install siding and roof covering	2.0
6. Install windows and external doors	1.0
7. Rough plumbing and electrical	1.0
8. Install sheet rock and floors	1.0
9. Install interior doors and finishes	1.0
10. Final plumbing and electrical	1.5
11. Inspection and occupancy	1.0
	12.0

Gantt Chart

This tool, invented by Henry Gantt, displays the activities such as those listed above in a series of bars. The length of the bar is proportional to the estimated time for the activity. Activities in sequence are shown end to end. Activities that occur simultaneously or can overlap with other activities are more clearly shown in this graphic presentation. Table 27.2 is a simplified example, using the same activities from Table 27.1.

Weaknesses of the Gantt chart are that it does not clearly identify task interrelationships. For example, Activity 6 above is shown starting during Activity 5; can it start earlier or later in relation to Activity 5? Also, this tool does not identify critical tasks or a critical path. If delayed, critical project elements will delay the entire project.

Critical Path Method

The Critical Path Method (CPM) was developed in 1956 at E. I. du Pont de Nemours. It was developed for a complex chemical plant project that had approximately 800 discrete activities.

Table 27.2 House Project Gantt Chart

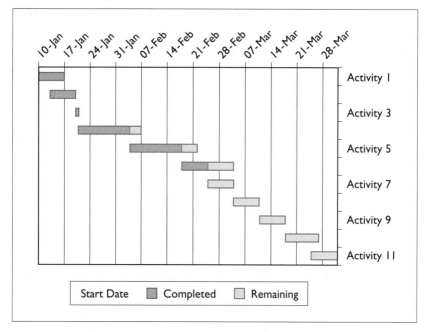

The CPM technique was enhanced by a program called Program Evaluation and Review Technique (PERT). This technique was invented by the U.S. Navy Bureau of Ordinance for the POLARIS missile development project. PERT added probability to the CPM tool. Each activity is assigned a set of probabilities for time to completion. Using stochastic computer analysis, this probability data can calculate probabilities associated with the entire project completion date.

With the CPM tool, we must again identify all activities and the time for each activity. Then we must identify the relationships between each activity. For example, Activity 3 and Activity 4 require Activity 2 to be completed. Another example is that Activity 10 requires Activity 3 to be completed, but the start date of Activity 3 is defined by delivery schedules, and this start date is unrelated to any other activity.

All the activities are presented graphically as a network. A network is two or more nodes coming together by ordered branches. The ordered branches are activities; activities can be tied together by the end of A coinciding with the start of B, or the end of B coinciding with the end of C, or many other combinations. The network can become quite complex with large projects.

The CPM tool also enables identification of the project's critical path. This sequence of tasks cannot be delayed if the project is to be completed on time. Noncritical paths include some "slack time" or "float" so that they can tolerate some delay without delaying the entire project.

The CPM tool offers a sophisticated mechanism to make sure the project is on schedule and to make adjustments by reassigning resources if the project gets behind schedule.

At this point you may be thinking, "Here we go again—I want to do engineering, but you are suggesting that I learn new tools, probability theory, and statistical analysis!" It's not all that bad; we do have computers to help us.

COMPUTER SOFTWARE

Numerous software programs can be used for project management. Some large organizations have developed their own to fit their operational and financial reporting requirements. The two best-selling programs are Primavera Project and Microsoft Project. In addition to using the concepts of CPM, these programs provide for cost tracking, human resource tracking, and multiple types of presentations. Find out which software program is used in your organization, and then try to learn it. You may even have a staff person in your organization who can help you get started with these remarkably user-friendly products.

These software programs have multiple features and capabilities. Versatile tools, they enable you to perform project monitoring and control, including the following tasks:

- Organize the project plan and think through the details of what must be done
- Schedule deadlines that must be met
- Schedule the tasks in the appropriate sequence
- Assign resources and costs to tasks and schedule tasks around the availability of resources
- Fine-tune the plan to satisfy time and budget constraints or to accommodate changes
- Provide links between elements of the project (tasks, resources, and assignments) and related project management documents in other applications

- Collaborate with other project stakeholders by reviewing the schedule and by notifying resources of their assignments
- Initiate and track discussions and resolutions of issues related to the project
- Prepare professional-looking reports to explain the project to stakeholders such as owners, top management, supervisors, workers, subcontractors, and the public
- Review the portfolio of all projects in the enterprise to analyze the impact of adding the new project on resource usage and cash flow
- Use portfolio modeling to optimize resource assignments across all enterprise projects
- Publish the project on a server for other project managers to access and for stakeholders to review via Internet browsers

When work begins on the project, you can use project management software to do the following:

- Track progress and analyze the evolving real schedule to see if it looks as though you will finish on time and within budget
- Notify resources of changes in their assignments and get progress reports on work that has been accomplished and that is yet to be done
- Revise the schedule to accommodate changes and unforeseen circumstances
- Try out different versions of proposed changes in a project, using "what-if" analysis, before making actual modifications to the plan
- Communicate with team members about changes in the schedule (even automatically notify those who are affected by changes) and solicit feedback about their progress
- Post automatically updated progress reports on the project server or on an Internet Web site or the company's intranet
- Produce final reports on the success of the project and evaluate problem areas for consideration in future projects

If this list looks impressive, even formidable, remember that the full capability of this software is needed only for very large projects. Early in your career, you will be involved in relatively small projects, or perhaps a small part of a larger project. For this project scope, you will only need the fundamental features of the project management software.

THE PROJECT PLAN

A project plan is needed for all projects, even for small projects. The project manager should prepare a formal project plan. This plan can be developed with assistance from project team members. Collaboration with future team members will help to ensure that no details are missed and proper documentation is completed. In creating the project plan, the project manager sets the course for the project. When completed, the project plan should be shared with all team members. Constant and full communication is essential for project success. The project plan provides a road map for project managers to follow. It also acts as the project manager's primary communications and control tool throughout the project.

What exactly is included in the project plan? The project plan is a set of living documents that can be expected to change over the life of the project. A common misconception is that the plan is just the project time line or the CPM diagrams. These are only components of the plan. The project plan is the major work product from the entire planning process, including all the planning documents. The project plan describes the purpose of the project and the expected results.

The project plan includes baselines and performance measures. These are the project's approved original specifications for scope, schedule, and cost. The project plan also includes documentation on how variances will be handled throughout the project.

Who is involved? The project plan should identify all stakeholders, those who have a vested interest in either the project or the project outcome. It should also identify all project team members. It is important for the project manager to get clarity and agreement on what work needs to be done by whom as well as which decisions each stakeholder will make.

What is the project's scope? The scope statement is arguably the most important document in the project plan. It is used to get common agreement among the stakeholders about the project definition. It is the basis for getting buy-in and agreement from the sponsor and other stakeholders and decreases the chances of miscommunication. This document will most likely grow and change with the life of the project. The project scope may be treated like a contract between the project manager and sponsor, one that can only be changed with sponsor approval.

Schedule and cost baselines. The project plan must identify activities and tasks needed to produce each of the identified deliverables. How detailed the task list needs to be depends on many factors, including the experience of the team, project risk and uncertainties, and ambiguity of specifications. This part includes the details of all the activities, their costs, their resource requirements, and their interrelationships.

Project management plans. The project plan should create the steps the team will take to manage variances to the plan. Different approval levels are usually needed for different types of changes. Not all new requests will result in changes to the scope, contracts, schedule, or budget, but a process is needed to study all new requests to determine their impact on the project.

THE PROJECT TEAM

In Chapter 26 on project procedures and team roles, we described several aspects of the engineering project team as well as the skills and professions that may be represented on the team. This section describes the project team for construction projects. No matter what type of engineering project you become involved with, team dynamics is a significant concern. This concern can also apply to small groups of engineers not working in a project team arrangement.

The effective project manager will want to select the most talented and experienced individuals for the project team. But these individuals must also be good team players. Some project teams are more effective than others, often not because of the complexity of the project. Rather, success comes because the team members work well together, work toward a shared goal, and make necessary changes and redirections in a coordinated and timely way. With this kind of working relationship, the project will move as rapidly as possible to a successful conclusion. Each team member also grows rapidly in experience and soon develops readiness for bigger and better assignments.

The effective project manager understands team dynamics and takes corrective actions to maintain harmonious relationships among team members. You can do your part by avoiding grandstanding and by avoiding being an obstructionist. Help the project manager to conduct smooth and effective project meetings. Be a contributor to solving problems; don't become the problem!

PROJECT POSTMORTEM

Despite effective project plans and good project management, problems will arise. Even successful projects will have portions that could have been done better. When the project is completed, the project team should hold a postmortem meeting. This is a lessons-learned exercise. All stakeholders in the project should have input to this review. The results of this postmortem review should be documented and used as resource material for future projects.

FOR MORE INFORMATION

- Project Management Institute: *www.pmi.org*
- David R. Pierce, Jr., *Project Planning and Control for Construction* (Kingston, Mass.: RS Means Company, 1988).

28

RISK MANAGEMENT

Risk management wasn't part of my early education and didn't impact my career as a young engineer. Perhaps that is because I started my career in the late 1950s, a time when litigation was not nearly as profuse as it is today. Furthermore, prior to the late 1950s, the professional liability insurance (PLI) industry was nonexistent. Not until about 1957 was the first PLI program created though the prodding of the American Institute of Architects (AIA) and the National Society of Professional Engineers (NSPE). Because of the lobbying of those two organizations on behalf of their members, who had no reasonable resource for errors and omissions insurance, the CNA Insurance Company PLI program was initiated. CNA is still the PLI program commended by both societies as of this writing.

When the insurance industry began offering PLI to engineers, it had no experience to draw upon in computing appropriate premiums or, more important, in learning the important aspects of defending claims and of developing risk management programs for their insured customers. The insurance industry's body of knowledge has dramatically expanded, and risk management training is one of the constants for the major carriers. They can draw upon nearly 50 years of legal defense experience and have learned much through monitoring outcomes in courts throughout the nation. Now training in risk management is an essential ingredient for managing an engineering firm.

What is risk? Many definitions are available. Some of those found in an Internet search are shown below:

- Source of danger (WordNet)
- Venture undertaken without regard to possible loss or injury (WordNet)
- The potential danger that threatens to harm or destroy an object, event, or person (Legal Encyclopedia)
- The potential harm that may arise from some present process or from some future event. In everyday usage, *risk* is often used synonymously with *probability,* but in professional risk assessments, risk combines the probability of a negative event occurring with how harmful that event will be (Wikipedia)
- The possibility of suffering harm or loss; danger; a factor, thing, element, or course involving uncertain danger; a hazard; the danger or probability of loss to an insurer; the amount that an insurance company stands to lose (American Heritage Dictionary)

WHAT RISK MEANS TO YOU

With a solid understanding of what risk means, it becomes obvious that we are surrounded by risks all the time. Risks arise in our personal life on a daily basis. They arise in our social and commercial activities as well. You learned, as a child, to "look both ways" as a risk management tool when preparing to cross a street. You later learned to read labels on prepared foods and especially on over-the-counter and prescription medications to avoid possible problems. You learned to dress appropriately for various weather conditions. You learned that while it was OK to jump up and shout at an athletic event, you should not do that in the middle of a classical music concert or in the classroom. Risk management is well ingrained in your day-to-day living pattern. You must apply the same type of learned behavior to your engineering career.

You may be wondering if you could incur any personal risk as a practicing engineer. That is a very good question and an important area of concern. The answer is yes—there is personal risk—but for an apprentice engineer, it is extremely low. Personal liability for errors and omissions, malpractice, or negligence cannot be shielded by a corporate structure. Even limited liability corporations, which are permitted by law in most states, do not furnish protection from personal claims in matters of malpractice or professional negligence, although there is personal protection from other kinds of liabilities, such as claims for unpaid corporate bills or leases, slip and fall incidents on the corporate premises, automobile accidents, and so forth.

Assuming you are or will be an employee of an engineering company, there is good news. Most companies carry PLI that provides coverage not only for the owners and principals of the firm but for all the employees. The coverage applies to any engineering work done on behalf of the company. It will pay for defense costs as well as any indemnity payment made to a claimant, less any deductible. You have a right to know if the firm you work for carries professional liability insurance. If it does not, you need to ask what the firm's policy is with respect to any malpractice or negligence claims made against employees. NSPE has adopted a "Position Statement on Disclosure of Professional Liability Insurance" that strongly encourages employers to disclose important information about their methods of protecting employees' interests. (That policy is included in the appendix.) A professionally ethical firm will indemnify its employees from expenses and personal losses that may arise out of work they perform for the firm.

RISK MANAGEMENT STRATEGIES

Risk management is the overall system by which a firm identifies and assesses its risks and by which it develops strategies to minimize or eliminate those risks. Let's look at the ways an engineering organization and its management can manage risk. We will look at some of the more prevalent risk generators and explore how to minimize or eliminate them. Fortunately, PLI carriers and attorneys who specialize in defense of claims against design professionals have already done a lot of the work necessary to identify risk for us. Many resources are available for your guidance. You might want to take a look at some of the Web sites that focus on risk management. Some good ones:

- Victor O. Schinnerer Company: *www.schinnerer.com*
- aeProNet: *www.aepronet.org*
- XLInsurance Company: *www.xldp.com/engineers/main.html*
- St. Paul Travelers Insurance Company: *www.stpaultravelers.com/ business_insurance/specialty/professionals/architects/index.html*

- Construction Risk.com: *www.constructionrisk.com*
- Ramco Insurance Company: *www.ramco-ins.com*

Once you have identified risks common to engineering practice, you need to develop strategies to deal with them. We will examine some common risks later on, but now let's look at some risk management strategies. They will fall into one of four categories:

1. Risk avoidance
2. Risk reduction
3. Risk retention
4. Risk transfer

Risk avoidance is simply not doing something that has risk associated with it. In the examples given earlier, it could mean that you choose not ever to cross any streets, or never to enter a room full of people. Of course, that would not be a realistic choice for risk avoidance. Although it is an exceptionally effective risk management tool, it is not a practical way to live or to run a business. Some risks are taken because the potential reward is worth the risk. You cross streets because you wish to reach a meaningful destination, and businesses take risks to earn revenue. In the case of engineering businesses, they not only want to make a profit, but they want to improve the living conditions of the people for whom they prepare project designs.

The strategy of *risk reduction* focuses on techniques for minimizing the probability of loss or injury. A personal example of risk reduction would be for you to travel with a raincoat, umbrella, and windbreaker so that you can handle inclement weather. You will be able to avoid discomfort and possible illness should unexpected weather conditions arrive on your trip. In a similar manner, engineering firms can look at potential threats of loss and take steps to operate in ways that will deter serious consequences from negative events.

The third risk management strategy is *risk retention*. That means that the firm will assess its capability to cover losses that may arise and will make a measured decision to accept risk with the intent of covering any expense that may arise out of an unanticipated event from its own reserves. If a firm does not carry PLI, it must have sufficient resources to cover such an eventuality and keep on running. If it does have PLI, the firm will pay a deductible for defense costs and any indemnity that may accrue, but the insurance will cover costs above that to the policy limits. Obviously, careful analysis is essential in determining the affordable retention level. To personalize this concept, if you have automobile or homeowner's insurance, the retention is the deductible amount in the policy declarations.

The fourth risk management strategy is *risk transfer.* Risk transfer means that someone else will be obligated, usually contractually, to assume some of your risk. A PLI policy is the most common way to transfer risk. In exchange for the payment of a premium, the insurance carrier will enter into contract (the policy) with the firm to cover defense costs and any indemnity payments to the policy limit, less any deductible amount (retention) agreed to by the firm. There are other forms of transference, such as indemnification and hold harmless agreements, but they are beyond the scope of this chapter.

AREAS OF POTENTIAL RISK

Some of the more prevalent areas of concern in risk management can be better managed if they can be anticipated in the contracting stage of a project. To assist in that, we will present a list of common areas of concern.

Client Selection

- Does the client have well-reasoned and appropriate expectations?
- Does the client have adequate financial strength?
- Does the client have a good reputation in its industry?
- Does the client have a history of litigation with other engineers?
- Does the client pay bills promptly and in full?
- Does the client seem to suggest that you need not worry about environmental or other regulatory issues in doing its work?
- Does the client suggest that the amount of work you need to do is minimal, compared to what you think it should be?

Not all of this information will be readily available. Business services, such as Dunn & Bradstreet (D&B), track and publish financial information about many firms. One can also check for complaints at the Web site of the better business bureau (*www.bbb.org*). Another resource is networking opportunities at technical and professional society meetings. You may find another engineer who has worked with the prospective client and who may be wiling to share an impression of the experience.

Contracts

One of the best means of practicing risk management is through the preparation of a good contract. We discuss contracts at length in Chapter 33, but their importance is so great that some of the information bears repeating here.

The best form of contract is a written one, and at the top of the line is the family of standard contracts produced by the Engineers Joint Contract Documents Committee (EJCDC). Those documents have contemplated virtually every important aspect. Next would be custom contracts written by the engineer, the client, or collaboratively. Risks are higher with custom contracts because there are countless areas of concern, some of which may not be addressed. We will list some of the critical areas below. It must be emphasized, however, that this book is intended to help you get oriented in your engineering career and, we hope, enhance your advancement. It is not to be considered a scholarly or authoritative treatise on contract law and should not be substituted for competent legal advice whenever a contract is written. Having said that, some key contract issues are the following:

- Is it written?
- Is it signed?
- Are the signers properly authorized to sign on behalf of their organization?
- Is the scope of project tightly defined?
- Is the scope of engineering services tightly defined?
- Is there a provision for renegotiation of contract should the client change the scope of the project?
- Does the contract have any wording that would elevate the engineer's service level above the legally required standard of care (equal to what would be rendered by a similar engineer in the same location)?
- Is it clear that opinions of construction costs (construction cost estimates) by the engineer are not guaranteed and that they will be made based upon recent experience and industry information? The final

construction cost will be the result of contractors' bids or negotiations and cannot be accurately determined by engineering estimates.

- Is there a person named for each party who will have the authority to act for that party?
- Is there a compensation and payment schedule?
- Is there a provision that the client will furnish application fees for applications for licenses, reviews, permits, and approvals made by the engineer on behalf of the client?
- Is there a provision for negotiating for extra work or for scope changes by the client?
- Is there a provision for the engineer to stop work without penalty in the event that invoices are unpaid after a specified period of time?
- Is there a provision that certifications requested by lenders or other parties that are not already shown in this contract will be the subjects of contract revision? Also, certifications by the engineer should be limited to subject matter under his or her control and within his or her knowledge. Some of the lenders and government agency certification forms go well beyond that level.
- Are any indemnification clauses in the contract overly broad?
- Does it say that ownership of documents will remain with the engineer (intellectual property)?
- Is there a condition establishing a first stage of dispute resolution before commencement of litigation, preferably mediation?

Subconsultants

- Is the subconsultant's scope of work clearly defined?
- Is there a written contract with the subconsultant?
- Is the subconsultant adequately insured, and have certificates of insurance (PLI and GLI) been furnished to the engineer?
- Is the engineer familiar with the subconsultant's reputation? Can references be furnished?
- Is the subconsultant's physical plant and staffing level adequate for the project?
- Is the subconsultant properly licensed, and does he or she have all the certifications that may be required for the assignment?
- Is the subconsultant acceptable to the client?

Construction Phase

- Will the engineer have construction administration duties? If not, who will interpret plans and specifications for the client? Will the engineer be consulted if change orders are requested?
- The engineer should not be responsible for contractor's ways, means, or methods of scheduling.
- The engineer should not be responsible for the construction site safety of any personnel, other than his or her own employees. (Some state laws deal with the engineer's responsibility for safety. Those laws are beyond the scope of this book but should be looked into on a state-by-state basis.)
- Does the engineer have a meticulously outlined policy of record retention?
- Does the engineer process shop drawing submittals in a timely fashion in accordance with a well-documented shop drawing processing policy?
- Does the engineer maintain a shop drawing log with all pertinent dates and processing actions?
- Has the engineer established formal lines of communication between his or her staff and the contractor?

Engineer's Office Operations and Staffing

- Is the project team composed of properly experienced personnel?
- Is the project manager properly trained and sufficiently experienced in team management and company policies?
- Is the project team structured in a manner to facilitate the exchange of ideas and communication?
- Is there an office operations manual that documents policies and procedures?
- Is there a formal policy for document retention?
- Is there a formal policy for management of electronic documents?
- Is there a formal policy for management of e-mail communication, both internally and externally?
- Is there a risk management training program for staff?
- Is there a person assigned to general risk management?

As you can see, the list of common areas of concern presented here is voluminous—and it is not nearly complete. The good news is that with experience, you won't need a list. Most of this will become second nature to you. As stated several times throughout the book, you should consider attending risk management seminars. If that is not possible, you should seek out a mentor in your organization and, at your mentor's convenience, you should solicit as much guidance as possible in this important area.

Early in this chapter, I gave some simple examples of risk management applications in your life before engineering. Had you not made those learned principles part of your daily life, you might have had some very bad experiences. At worst, you might not even be here. Risk management in your engineering life is as important, and it can become just as natural if you give it a chance. By the way, risk management is required in all businesses and professions so don't be discouraged about your choice of profession. Engineering is one of the most important professions, one of the proudest, and one of the most respected.

29

LEGAL CONCERNS

OVERVIEW

Engineering is a profession that has a significant influence over society in just about every area of its citizens' lives. As is the case for all professions, engineering is controlled by statutes and regulations in every U.S. state and territory (jurisdiction).

U.S. jurisdictions have been delegated the power to protect the health, welfare, safety, and property of the public by the federal government. Under that power, known as police power, professionals are required to prove that they have met certain requirements as evidence that they have achieved at least the minimum accepted level of competence to serve the public. All U.S. jurisdictions will require an engineer to prove that he has met the educational, experience, and training levels specified by law and, after satisfying those criteria, that he can pass an examination to prove that the necessary knowledge has been imparted to the engineer.

Upon passing the examination, the candidate will be granted a license to practice as a professional engineer. Once that status is attained, certain privileges will accrue to the candidate along with certain obligations and duties. Licensure is discussed more thoroughly in Chapter 17, but it is essential to understand that only a licensed professional engineer can legally be in responsible charge of an engineering project or offer services directly to the public. Licensure imposes certain rigorous legal standards upon the licensee. There are, however, exemptions from the requirement for licensure for those who will

- work in industry and furnish engineering services only to their employers;
- work for certain government agencies (depends upon jurisdiction); or
- work under the direct supervision of a licensed professional engineer, with no individual need to furnish direct service to the public.

For purposes of this chapter, we will assume that most readers will go on to become licensed professional engineers. We sincerely hope that reading this book will inspire you to seek licensure.

Engineers, licensed or not, will be expected to practice legally and observe appropriate legal constraints. Legal practice means that one will observe each and every law that relates to one's daily activities and conduct. This can be a tall order as you will see. The licensure law imposes specific constraints, but it is far from the only legal area of concern to engineers. A plethora of laws exists, and new and/or amended laws must also be observed. Therefore, legal compliance presents a challenge.

HOW LAWS AND REGULATIONS ARE CREATED

Laws are rules of conduct that describe and control permissible and/or prohibited conduct of people and organizations. They provide for penalties against those who violate the prescribed limits of acceptable conduct. Laws are administered through the court system.

Within the overall body of law, the highest and most inviolable is constitutional law, which grants the power to legislate and to enforce legislation to other bodies of government. As enabled by the United States Constitution, every state has its own state constitution. State laws are called statutes. Laws at lower levels, such as counties and municipalities, are usually called ordinances. Some laws set down the general outline of what the legislators are trying to accomplish and include a section delegating power to an agency (or creating one) to write regulations detailing the intent of the legislature.

Regulations set forth detailed rules of conduct in a wide variety of activities. They are drafted by agencies that have been empowered to do so by the legislature. The regulatory process normally includes public review and comment before draft regulations are finalized. Once public input has been considered, the regulations are adopted and promulgated.

Regulations are subject to periodic revision without legislative oversight. Because regulations can be amended with relative ease and frequency, it is important to monitor those that are particularly important to one's day-to-day practice. Regulations carry the force of law, and practicing engineers need to be aware of and compliant with them. The engineering community and others in the business community have frequently complained about the unlimited power exercised by agencies framing regulations.

SOME AREAS OF STATUTORY AND REGULATORY CONTROL

Engineers are expected to practice legally, observing the appropriate and applicable legal constraints. As previously stated, legal practice means observing each and every statute and regulation that relates to your conduct. The statute and regulations pertinent to the practice of engineering in one's state(s) of registration are of major importance. They should be reviewed on an ongoing basis so that the engineer can remain in touch with their constraints, particularly as they relate to potential misconduct. But the licensure laws are not the only place one needs to look for controlling law.

There are many laws that are applicable to the practice of engineering. A partial list of areas of concern is presented below. Some of them are federally created, but most are state created. Some of them apply only in some states, not all.

- Environmental
- Land use (zoning, planning, application, and permitting procedures)
- Occupational Safety and Health Act (OSHA)
- Employment/labor
- Statute of limitation
- Statute of repose
- Certificate (affidavit of merit)
- Americans with Disabilities Act

- Patent
- Copyright
- Qualifications-based selection (QBS)
- Federal acquisition regulations (FAR)
- Design-build
- Business (business forms, contracts, etc.)
- Sole source workers' compensation
- Joint and several liability
- Products liability

As stated, the list is not complete but is intended to convey the broad scope of legal constraints that can impact the practice of engineering. Many of the listed areas may not apply to your career choice or may apply only on occasion. Most employers have someone who is charged with monitoring laws and regulations to stay abreast of and manage practice with the boundaries of conduct prescribed by law. Of course, if you aspire to become a manager as your career advances, you may eventually find that the monitoring of laws and regulations is your task. It is definitely not recommended that you get carried away and begin to make decisions based upon your interpretation of the law, or that you give legal advice. That is the stuff for which lawyers exist. Without familiarity with the legal landscape, however, you can become lost; possibly unaware of areas in which legal advice is essential.

To provide a better understanding of the legal areas listed above, we've described in further detail the functions of key areas of law.

ENVIRONMENTAL LAWS

This class of laws is intended to provide a cleaner and safer environment for our citizens. They are always crafted by either the U.S. Congress or jurisdiction legislatures. They will normally empower an agency to craft and promulgate regulations that give detail to the intent of the legislature. Some of the areas that are controlled by environmental laws and regulations are

- storm water management
- freshwater wetlands protection
- flood plain management

- air quality
- sanitary sewage collection and treatment
- water diversion (wells)
- waste management
- coastal area protection
- recycling
- hazardous material management

In the engineering practice areas of industry, government, and private practice, environmental laws impact engineering design and decision making. Some of the most highly publicized environmental contaminations have arisen out of manufacturing plants that indiscriminately disposed of hazardous materials within their manufacturing facilities, albeit the damage was done long before the environmental consequences could have been predicted. Military installations as well have created significantly hazardous contaminants from dumped munitions, petroleum products, and other materials now known to be a threat to the groundwater supply and to anyone who should live near or on top of the contaminated soils. Engineers in industry need to be sensitive to advising their employers on environmental issues. Those in government are often engaged in ferreting out such problems and enforcing the laws; those in private practice are often engaged in devising the solution for cleaning up the problems sites.

OSHA LAWS AND REGULATIONS

The federal government enacted the Occupational Safety and Health Act with the intent of providing all employees with a safe and healthy workplace. It is required of almost every employer to do that. For engineering firms engaged in construction administration, observation, inspection, or stakeout, recent developments in the law have made it critically important to be on the alert for hazards to construction workers. This area of law is still developing.

Some engineering firms feel that they should not have any duty to observe the safety of a construction worker who is under the control of a contractor. That is actually a valid point, and construction contracts almost always place the safety of construction workers in the hands of their employers. Some court cases, however, have gone beyond the contract language, imposing a duty upon engineers to act upon any imminently dangerous conditions that are apparent to them at a construction site. This is a developing body of law (state-by-state), and an engineer must know what his firm's policy is with respect to construction site safety of construction workers. That said, an engineer's employer definitely has a legal duty to be cognizant of the safety of the employees of the firm, both in the office and at a construction site.

STATUTE OF REPOSE

Statute of repose is a law that bars actions against engineers and other design professionals a certain number of years following completion of professional services or after substantial completion of the construction project. The underlying principle that supports the passage of such a law is that the evidence that would ordinarily be relied upon, including testimony of witnesses, becomes "stale" over time. Witnesses may disappear or die, and even if they are available, memories become clouded. The statute of repose exists in all jurisdictions, although the time limitation varies. Being aware of the statute of repose will aid engineers and firms in making rational decisions about retention of records and about making an effort to preserve the record of the triggering event that starts the beginning of the appropriate period.

PRODUCT LIABILITY

In Chapter 12, we discussed professional liability. Professional liability of engineers is measured by standard of care (i.e., even if damages arise out of an engineering project, the courts do not hold the engineer to a standard of perfection). With respect to products, however, the concept applied by the courts is that when products are offered to the public for sale, the manufacturer is warranting that the product is suitable and safe for its intended use. In such circumstances, the courts will apply strict liability. It is deemed irrelevant that the manufacturer or vendor of the product did not act recklessly or negligently. In fact, the manufacturer or vendor could have been totally unaware of the potential injury caused by the product.

Many jurisdictions have enacted stringent product liability laws. There is no federal product liability law, but the U.S. Department of Commerce has issued a guide called the "Model Uniform Products Liability Act" for voluntary use by jurisdictions. This application of law will most likely have more meaning to manufacturers and vendors than to the engineers who work on design, but an engineer working on product design should have the reassurance of the employer that, in the unlikely event that the engineer is named personally in a product liability action, the engineer will incur no personal costs for defense or for indemnification (payment to the plaintiff).

SOLE SOURCE WORKERS' COMPENSATION

When workers' compensation laws were enacted in the early 20th century, employers became protected from claims of negligence by their employees who became injured on the job. The insurance furnished by the employers covers lost wages and medical costs, but should an injured employee wish to bring a claim for pain and suffering, or other damages, the employer is protected.

As a result, injured employees scan the horizon for third parties who may have played a role in the accident that injured them. Among the obvious targets are the engineers who prepared the plans and specifications for construction projects, where many injuries do occur. This phenomenon is enhanced by the fact that most engineers carry professional liability insurance, giving the injured party a "deep pocket" to pursue. A number of jurisdictions, about a third of them, have adopted some form of protection for design professionals, furnishing immunity for engineers against such suits under most circumstances. The majority of jurisdictions, however, do not provide such immunity.

CERTIFICATE (AFFIDAVIT OF MERIT)

This type of law is not predominant, existing in only about one quarter of the jurisdictions. The form varies from jurisdiction to jurisdiction, but it requires someone who wants to sue an engineer to get the opinion of an expert that there is good reason to pursue the litigation. The expert who issues the affidavit is normally required to be licensed and practicing for a reasonable period of time in the same field as the defendant engineer.

The intent of this type of legislation is to impede the filing of frivolous (without merit) lawsuits. It is believed that counsel for a potential plaintiff will be less likely to file if a properly experienced and credible expert is not willing to file the necessary opinion. There is no way to measure the number of lawsuits that are not filed because of this requirement, but most experts believe it does achieve the intended effect most of the time.

QUALIFICATIONS-BASED SELECTION (QBS)

This type of law requires some jurisdictions, as well as some counties, municipalities, agencies, and universities, to procure design services without using a bidding process. Instead, the owner will issue a request for qualifications (RFQ) for a project and review the qualifications of submitting firms. Then the owner will rank the respondents and begin the negotiation stage by interviewing the top-ranked firm. If the parties cannot reach a satisfactory fee arrangement, the owner will repeat the process with the second-ranked firm, and so on. The actual process is quite detailed, with good guidelines produced by the leading engineering organizations, as well as the American Bar Association and the American Public Works Association.

Congress passed the granddaddy QBS bill as Public Law 92-582 (Brooks Bill) in 1972. It confirms that QBS is in the nation's best interest in federal procurement on civilian agency projects. The federal requirement is still in place, and the rules that flowed from it are known as the Federal Acquisition Regulations.

LAND USE (ZONING, PLANNING, APPLICATION, AND PERMITTING PROCEDURES)

For engineers who work in private practice or in government, they likely will become involved in land development applications at some point. In private practice, engineers will work to plan layouts of residential, commercial, or industrial developments on raw tracts of land. The environmental laws discussed above will be a big part of the consideration in doing the planning and, eventually, the detailed design of the proposed developments.

To do that sort of work, the engineer will apply technical skills related to storm water management, collection and disposal of sanitary sewage, water supply and distribution, roadway alignment (horizontal and vertical), earth grading, landscape design, structures, utility access, and more. Client relations will be very important in this kind of work, as well as coordination with a project team, usually consisting of the owner-developer, an architect, and a land surveyor. Others on the team may include specialists such as, but not limited to, a structural engineer, a geotechnical engineer, an urban planner, an environmentalist, and a land planner.

To produce a plan that has a good chance of obtaining the necessary local, county, and state approvals and permits, the engineer must be familiar with the local zoning ordinances that will control the permissible uses and lot sizes, as well as a multitude of other details. The engineer will also have to master the application requirements of the various reviewing authorities, which can be many, with respect to plan details, submission schedules, application fees, design standards, and more.

It is not unusual for a project to have multiple agencies with jurisdiction. A typical project might include the municipal land use board; the departments of health, fire, and police; the shade tree commission; and the city engineering department. At the county level, one may interface with the engineering department, road department, planning board, and environmental commission. At the state level, there may be the soil erosion control district, the department of environmental protections (perhaps several divisions, such as wetlands, stream encroachment, hazardous waste, sanitary sewers, water distribution), and sometimes review by agencies with authority over condominium development. There may also be a sewage or water authority to satisfy, usually autonomous agencies set up by local government. In some jurisdictions, utility owners, such as electric, gas, cable, telephone, and water companies, may have authority and review rights over the design. They have their own standards, rules, and regulations and, often, engineering departments who will conduct reviews.

When the project is submitted with an application for approval, the engineer will be called upon to attend meetings with the appropriate professionals representing the authorities and to appear at public meetings to testify and to defend the application on behalf of the applicant.

For engineers in government, particularly at the local or county level, the review responsibility for submissions prepared as above will fall to you. Government engineers will be challenged to review the plans, reports, specifications, and studies submitted by the applicant's engineer for completeness (does it have everything required by ordinance and law?) and for technical proficiency. They will be called upon to write a report to the appropriate board of your organization and to attend meetings to discuss the projects with the applicant, the applicant's project team, and the public.

As difficult as this may all sound, I spent nearly four decades in that kind of environment and found it not only challenging but also personally gratifying. Staying abreast of the laws and regulations was a very big challenge. It was necessary to monitor changes in many ways, including (for me) subscribing to the *New Jersey Register*, which published all proposed regulations and proposed amendments to regulations for New Jersey. Local newspapers publish proposed local ordinances and amendments to ordinances. It can't be overstated that the monitoring of laws and regulations is a critically important element of competent practice.

One more resource is available to engineers who wish to stay abreast of legal issues and, possibly, influence legislation or rulemaking. The professional and technical societies and associations all make an effort to stay abreast of developing laws and rules that will affect their members. Most of them monitor the legislature and the regulatory agencies and notify their members when laws and ordinances are up for amendment or when new ones are being generated. In matters having significant positive or negative impact upon their members, they lobby on behalf of the interest of the membership and notify the membership of any opportunities to participate, possibly as committee or task force members, or though letter writing, e-mailing, or calling appropriate people to lobby for the desired outcome. Every practicing engineer should consider membership in at least one professional or technical association for that reason alone, although there are many more reasons to belong to such associations.

CONCLUDING THOUGHT

Some see the law as a set of restrictions. Perhaps, in a sense, that is the case. Imagine a civilization without laws, however. Our daily lives could become really chaotic. All of us have had times when the law, and especially regulations, have given us grief, seeming to obstruct more than help. However, another view of laws and regulations is that they furnish a reference, a guide to legally acceptable practice. Complaining among ourselves about laws is a useless waste of energy and time. Perhaps some joint effort through an organization or communication with our legislators may have an effect. Short of that, however, the best thing to do is to understand what laws require of us and make every effort to comply. That is especially true for a profession that has such a significant and often enduring impact upon the public. An engineer will be best served by learning about and observing the law.

30

LAWS, REGULATIONS, CODES, AND STANDARDS

Throughout this book, you are reading about the professional, ethical, and legal responsibilities of engineers. In this chapter, we will focus upon the topic of laws, regulations, codes, and standards. You need to know how they arise and the distinctions among them. No matter what segment of engineering you have chosen to call your professional home, laws, regulations, codes, and standards will apply to you or, at the very least, to your employer.

One note of caution: Many engineers who work in industry may hold that these areas are not your concern, and in many cases that may be true. However, if an industrial employer (or any other employer, for that matter) sees profit as the paramount goal, and if conforming to the requirements of laws, regulations, codes, and standards is an impediment to profit, your management may be inclined to "wink the eye." We can only hope that, through the various professional coaching and mentoring guidelines presented in this book, you will at least consider the implications of ignoring legalities. Laws and regulations almost always have penalties for violation attached to them, and some codes and standards, when adopted by law, also carry such consequences.

TYPES OF LAW

Laws define the permissible conduct of people and organizations. Laws provide for penalties against those who fail to act within the prescribed rules of conduct. Laws are administered through a system of courts, collectively known as a legal system.

Within the overall body of law, there is a hierarchy. Our highest and most sacrosanct body of law is the U. S. Constitution. Other bodies of law important to the engineering community are civil law and common law. There are also regulations, codes, and standards that govern the work of engineers.

Civil Law

Civil law is a system of rules of conduct enacted at the federal, state, or local level (ordinances) by elected officials. Occasionally, law may be created by the chief elected official (president or governor) through the issuance of an executive order.

Some laws are categorized as "enabling legislation" that set forth the general intent of the legislature but designate another body (commission, board, government agency, etc.) to implement the law through rule making. That rule-making body will then draft regulations. Boards of registration for professional engineers are an example of that process.

Common Law

Common law plays a significant role in shaping the body of law. A simplified description of common law is that it is based upon unwritten laws, largely comprised of judicial decisions that may be relied upon by the legal community and the courts as binding precedents. Appellate division and supreme court decisions are often recited in legal arguments in the courtroom as supportive of the arguments presented. In a June 23, 2005, decision, the U.S. Supreme Court ruled that government could use eminent domain to seize private property for use by developers if a public benefit was expected. That set the real estate development world on its heels, some say opening the door to political abuses. That decision has the force of law.

Regulations

Regulations comprise a body of detailed rules controlling conduct in a wide variety of activities. Agencies of the government, empowered to do so through enabling legislation, draft regulations. There is provision for public review and comment after regulations have been drafted. After public input has been considered, the regulations may be revised, or they may be finalized and promulgated. Regulations, unlike laws, can be periodically revised without legislative oversight. Engineers and others in the business community have often voiced displeasure with the regulatory process. They are troubled by the seemingly unlimited power wielded by agencies that are empowered to draft regulations. Nevertheless, regulations carry the force of law, and those that are applicable to the work at hand must be on every engineer's radar screen.

Codes

A *code* refers to a set of standards that has been officially adopted by one or several governmental bodies. Once adopted, a code takes on the force of law.

Standards

A *standard* refers to technical specifications drafted as guidelines for manufacturers, installers, designers, and for constructors of facilities. The scope and breadth of standards is virtually limitless. Standards are the output of groups of experts who are organized for the express purpose of producing such standards. For a list of various standards-producing organizations, visit *www.library.umaine.edu/science/Standard.htm.*

The standards-producing experts operate predominantly within the framework of a large technical organization, such as the American Society of Mechanical Engineers (ASME), the Institute of Electrical and Electronics Engineers (IEEE), the National Fire Protection Association (NFPA), the American Society of Civil Engineers (ASCE), the American National Standards Institute (ANSI), the American Society for Testing and Materials (ASTM,) and so on. Standards do not carry the force of law unless they become codified through adoption by a governmental agency. That is why standards are often termed "voluntary" when the engineering community uses them. Voluntary standards provide the opportunity for engineers to rely upon the expertise of a select group.

THE ENGINEER'S LEGAL STANDARD

Chapter 12 provides an in-depth discussion of professional liability for engineers. It is important, however, to preface a discussion of laws, codes, standards, and regulations with a brief discussion of the legal standard applied to the examination of alleged negligence of an engineer.

In the United States, the universal standard by which negligence of an engineer is measured is found in common law, where it is well established that an engineer is required to act with that degree of competence to be expected of an average engineer practicing under substantially similar conditions, at the same time and in the same location as the engineer under review. The question of compliance with that standard almost always requires expert testimony by someone qualified by the court to evaluate the conduct of the defendant engineer in the matter at hand. One area the expert will examine is the engineer's compliance with duty owed to the party bringing the complaint. An engineer's duty can arise from a variety of sources, including

- contracts;
- laws, codes, standards, and regulations;
- ethical and professional responsibilities; and
- common law.

The focus of this discussion is laws, codes, standards, and regulations. The engineer is duty-bound to be aware of, and to comply with, applicable laws, regulations, codes, and standards. In reality, however, to be fully cognizant and fluent with all laws, regulations, codes, and standards relevant to the project at hand is often impossible. Clients may submit contract language that would require the engineer to produce a design that is compliant with "all laws, regulations, standards, and codes," but such a condition is onerous, eminently unfair, and nearly impossible to meet. An engineer should vigorously resist such language.

Nonetheless, an engineer is duty-bound to the client and to the public to make a best effort to ferret out and comply with all applicable laws, regulations, standards, and codes that are reasonably attainable and that would be utilized by the average engineer practicing under substantially similar conditions, at the same time and in the same place.

All practicing engineers with a modest amount of experience, and experience particularly related to environmental permitting issues, are aware that compliance with *all* readily available laws, regulations, standards, and codes may be literally impossible, because at least some of them will be in conflict with other legal requirements. That is the consequence of well-meaning but narrowly focused multiple agencies and/or branches of government making uncoordinated laws and/or regulations. Even the best effort of an engineer to comply with all laws, regulations, standards, and codes can meet with failure. Failure may result in a claim of breach of contract that requires compliance with "all laws, regulations, standards, and codes." Legal compliance is a challenge for the best of us.

Furthermore, laws, regulations, standards, and codes may be subject to change, cancellation, or replacement with new ones. The engineer should not be penalized for such eventualities. It is prudent to address the possibility of such changes in the professional services agreement/contract.

Don't ever forget, an engineer owes a duty to the client (and to the public) to ascertain, to the degree reasonably possible, what laws, regulations, standards, and codes apply to a project under design.

THREE ESSENTIAL LEGAL PARAMETERS

Although the number of laws, regulations, standards, and codes is enormous, it is appropriate to mention that one law, one set of regulations, and one code are of singular significance for every licensed professional engineer. These legal parameters must be understood and monitored for changes that may occur over time. They are

- the licensure statute;
- the regulations defining conduct of a licensed professional engineer; and
- the code of ethics of at least one recognized engineering association.

The statute establishing the professional engineering license in your state(s) of practice is extremely important. It sets forth the intent of the legislature with respect to minimum qualifications and methods of attaining and maintaining licensure. Inevitably, the licensure statute will consider the health, safety, and welfare of the public as the underlying foundation for the law. (The licensure process is covered in depth in Chapter 17.)

The statute usually empowers a licensing or registration board to create and promulgate details governing the practice of a licensed engineer. Specifications of the code of conduct of practice are set forth within the regulations. The importance of obtaining and studying these important controlling documents cannot be overstated.

In addition to knowing the provisions of the licensure statute and the regulations, you should obtain and become familiar with the code of ethics of the associations to which you belong. If you are not a member of an association, then you should strongly consider joining one. Failing that, it is a safe move to obtain and study the Code of Ethics of the National Society of Professional Engineers. (A copy of the current version is included in the appendix.) Engineers and the courts widely recognize it as a set of guidelines for ethical practice. It should be noted that the NSPE Code of Ethics is upgraded from time to time to address developing issues. In 2006, for example, it was amended to say that engineers should strive to use sustainable development in their designs. The NSPE Code of Ethics can be obtained for no charge at *www.nspe.org/ethics/eh1-code.asp.*

LAWS AFFECTING ENGINEERS AND ENGINEERING COMPANIES

Listing every law affecting engineers and engineering companies at the federal and state level would be virtually impossible. In an effort to demonstrate the broad coverage of existing laws, the following is presented for your general guidance. Be advised, however, that this book is not intended to be a substitute for sound legal guidance from an attorney.

There are almost limitless laws that can impact both technical and business practice. In an effort to give at least a modest amount of insight and guidance, the following list has been compiled. Though not complete or exhaustive, for those who never considered the breadth of legal constraints on practice, the list should be revealing. To the extent possible, items are indicated as federal (F), state (S), or local (L).

- *Patent law (F) (S)*. This subject is covered in depth in Chapter 31.
- *Copyright, with a focus on intellectual property (F)*. This subject is covered in Chapter 31.
- *Brooks Bill, Public Law 92-582 (F)*. This law established that qualification-based selection (QBS) of design firms was in the nation's best interest. (QBS is covered in Chapter 32.)
- *Federal Acquisition Regulations, or FAR (F)*. This set of regulations applies to acquisition procedures to be followed by the federal government and its agencies. Procurement procedures (QBS) for architect-engineer (A/E) services are covered in subpart 36.6.
- *QBS procurement laws (S)*. QBS procurement of A/E services has been adopted by almost all states (47). Research on these laws must be done on a state-by-state basis. Although QBS may be adopted for state procurement, be aware that counties or municipalities within the state may not mandate it; research of each unit's procedure is necessary. Other subdivisions of that government, such as transportation or utility authorities, universities, and so forth, may subscribe to QBS acquisition procedures.
- *Design-build laws (S)*. These laws must be researched on a state-by-state basis. Assistance can be found at the Design-Build Institute of America's (DBIA) Web site, *www.dbia.org.*
- *Environmental laws (F), (S), (L)*. There is a nearly unlimited quagmire of environmental legislation at all levels covering air quality, water quality, storm water management, stream encroachment, flood plain management, wetlands protection, soil erosion and sediment control, hazardous waste management, recycling, and more. Environmental law has a major influence upon the practice of engineering in the United States, and there is no reason to think that it will diminish in the future. Such legislation is undeniably motivated by lofty purpose, but some engineers feel that protecting the environment has gone too far. While that debate continues, this body of law has produced some prime examples of contradictory requirements that would make it virtually impossible for anyone, including engineers, to comply with all provisions.

Some major environmental and other laws at the federal level are these:

- National Environmental Policy Act of 1969 (NEPA)
- Chemical Safety Information, Site Security, and Fuels Regulatory Relief Act
- The Clean Air Act (CAA)
- The Clean Water Act (CWA)
- Comprehensive Environmental Response Compensation, and Liability Act (CERCLA or Superfund)
- Emergency Planning and Community Right to Know Act (EPCRA)
- Endangered Species Act (ESA)
- Federal Insecticide, Fungicide, and Rodenticide Act (FIFRA)
- Freedom of Information Act (FOIA)
- Oil Pollution Act of 1990 (OPA)
- Pollution Prevention Act (PPA)
- Resource Conservation and Recovery Act (RCRA)
- Safe Drinking Water Act (SDWA)
- Superfund Amendments and Reauthorization Act (SARA)
- Toxic Substances and Control Act (TSCA)

- *Land use laws (S), (L).* These laws, often associated with environmental laws, set forth the required procedures for development of land for residential, commercial, and industrial purposes within communities across the nation. State law may set forth the requirements and constraints that must be followed by political subdivisions within the state; local ordinances are required to comply with state laws. Land use laws set out a detailed set of rules for site plans or subdivisions in each community. Local ordinances cover such things as zoning, procedures for submissions for land use development reviews, specifications for content of application packages, including the required details to be shown on plans drawn by the engineer, storm water management plan requirements, minimum design standards, application fee schedules, and more. In some areas, county rules and regulations will have an effect as well.

- *Occupational Safety and Health Act (F).* This law requires all employers, including engineering firms, to comply with the standards developed by the Occupational Safety and Health Administration (OSHA) to provide a safe and healthful workplace for all employees. Many engineers think in terms of the safety of construction workers at the construction site where their plans are being implemented. Construction work presents a serious risk management issue, but engineering firms should not lose sight of the fact that their own workforces must be afforded the same protection required for construction workers employed by contractors.

- *OSHA regulations (F)*. The OSHA regulations most relevant to engineers are found in the Code of Federal Regulations at 29 CFR 1910 (general industry) and 29 CFR 1926 (construction). The regulations set out detailed standards for workplace environments. As stated above, every engineering employer has a duty to see to it that its employees are afforded a safe and healthful workplace. When an engineering employer's workforce (including field personnel at construction sites) is properly trained in workplace safety, then it can better identify hazardous working conditions that may affect construction workers.
- *Americans with Disabilities Act, or ADA (F)*
- *Employment Laws (F)*. To list just a few:
  - Contract Work Hours and Safety Standards Act
  - Copeland Antikickback Act
  - Davis Bacon and related acts
  - Employer Retirement Income and Security Act (ERISA)
  - Fair Labor Standards Act
  - Immigration and Naturalization Act
  - Whistleblower Protection Act (whistle blowing is discussed in Chapter 11)
- *Statute of limitations (S)*
- *Statute of repose (S)**
- *Sole source workers' compensation (S)**
- *Certificate or affidavit of merit (S)**
- *Joint and several liability (S)**

Information on the four state laws marked with * can be found in a publication of NSPE entitled "A State-by-State Summary of Liability Laws Affecting the Practice of Engineering." That publication, prepared by the Professional Liability Committee of NSPE Professional Engineers in Private Practice, tabulates which states have enacted legislation in these four key areas. The publication can be purchased from the NSPE Web site through the products and services page.

SPECIAL CIRCUMSTANCE: NATIVE AMERICAN LANDS

Entering into contracts with Native American tribes for design or construction services presents special challenges. Native American tribes are independent entities. They have special rights bestowed upon them by the federal government. In general, contract disputes between Native American tribes and those providing contractual goods or services are adjudicated under the affected tribe's law. Also, contracts between Native American tribes and others generally require the approval of the U.S. Bureau of Indian Affairs. Anyone considering service relationships with Native American tribes should seek competent counsel from an attorney experienced in Native American contracting issues and rely upon his or her firm's professional liability insurance carrier for guidance.

CONCLUDING THOUGHT

You may be somewhat perplexed by the long list of laws presented above. Let me assure you that you will not be expected to learn all of them. Most of the time, it becomes apparent when one or more will come into play. Our intention is to sensitize you to circumstances when you need to consult with someone about applicable law. You cannot engineer "by the seat of your pants," a term for how the old-time barnstorming pilots flew their planes. Today no pilot can think about flying that way unless he is a wild risk taker. Pilots need thorough training, good instruments, a reliable engine, and a solid flight plan. Let your flight plan carry you to success. This book is just one tool you can use to get where you want to go.

31

PROTECTING INTELLECTUAL PROPERTY

Patents, Copyrights, Trademarks, and Trade Secrets

Throughout your career, you will be asked to create new and innovative designs or products that you will want to protect. Therefore, it is essential for you to have a basic understanding of what intellectual property is and how it is protected. Especially those engineers who will work in industry or are doing research work at a university should be aware of the general principles regarding intellectual property protection. In this chapter, we look at some of the technicalities of patents, copyrights, trademarks, and trade secrets.

All four elements—patents, copyrights, trademarks, and trade secrets— are forms of intellectual property.

PATENTS

Patents are grants issued by the U.S. Patent and Trademark Office (USPTO) to inventors. A patent for an invention gives its inventor the right to exclude others from making, using, selling, or offering to sell the invention in the United States or its territories or possessions.

Patent law in the United States dates back to 1646, when the General Court of Massachusetts issued the first patent, which protected a mill for manufacturing scythes. In 1790, President Washington signed into law the bill that set into motion our modern U.S. patent system, affording protection to American inventors from unauthorized application of their inventions without just compensation.

Domestically, patents are now granted exclusively by the U.S. government. The authority for patents comes from the U.S. Constitution, article 1, section 8, which states: "The Congress shall have power to promote the progress of science and useful arts by securing for limited time to authors and inventors the exclusive right to their respective writings and discoveries." The protection arising from the granting of a patent, however, does not run forever. The patent system is administered by the United States Patent and Trademark Office (USPTO). The three categories of patents are the following:

1. *Utility patents.* Cover any new and useful process, machine, manufacture, composition of matter, or any new and useful improvement thereof. Endure for 20 years from the date of filing.
2. *Plant patents.* Granted on any distinct and new variety of asexually produced plant. Have a life of 20 years from the date of filing.
3. *Design patents.* Granted on any new, original, and ornamental design for an article of manufacture. Endure for 14 years from the date of filing.

Life spans for patents can be extended only by an act of Congress, although some special rules apply to pharmaceutical patents because of the extensive time consumed during government-mandated testing periods.

To obtain a patent, the inventor must file an application with the USPTO. There are certain stringent requirements as to content and style of the application, so it is strongly recommended, but not required, that a competent and experienced patent attorney assist with the application process.

Corporations or other business entities that may have contractual rights to an invention generally cannot sign an application unless they own the invention and the inventor refuses to sign. Otherwise, only the inventor can sign the application for a patent grant. A filing fee must be submitted with the application, although the amount will be lower for an inventor or a small company than for a larger organization.

With respect to contractual rights assigned to the employer, you may want to be aware of this possibility. Sometimes new hires in engineering positions will be given a contract to sign but not pay attention to the details. Contracts containing patent clauses may be brought to an attorney for review.

If nonnegotiable and a condition of employment, then you may have no alternative but to sign the contract.

Before an application for a patent grant is filed, inventors and companies should conduct a patent search to be reasonably certain that the invention is not similar to a previously patented invention. The search can be conducted by a patent search firm or by the inventor. Many sources can be searched, including but not limited to

- USPTO weekly patent summaries
- online databases of patents
- the USPTO offices in Alexandria, Virginia
- state-based USPTO depositories; the location and number must be determined on a state-by-state basis; information is available at *patents.uspto.gov/web/offices/ac/ido/ptdl*
- the USPTO database containing all patents published since 1976 and patent applications published since March 15, 2001; the Web site is *patents.uspto.gov/web/menu/search.html*
- Also available at this Web site are graphical copies of all pre-1976 patents dating back to 1790

When the search is satisfactorily completed, a patent attorney will be able to form an opinion as to the likelihood of success of the patent grant application.

After an application is filed with the USPTO, it will be assigned to an examiner in an appropriate technology group. The process may consume up to four years, with two years being about average. Often the examiner will raise questions about the similarity of the invention to previous inventions, leading to interaction between the examiner and the inventor or, more likely, with the inventor's attorney. The inventor or his or her attorney will try to overcome the examiner's reservations by explaining why the prior patents are not pertinent. Eventually, the application will either be denied or a grant will be given.

The ownership of an invention will normally be with the inventor. Exceptions to this principle are for inventors contractually employed by companies that hired them expressly to invent, or inventors who have contractually assigned any inventions to their employer. In that circumstance, the company will own the invention. If an employee was not hired "to invent" but does so using the employer's time and resources, then the company may have "shop rights" to the invention. In that circumstance, the employer can use the invention without any additional compensation to the inventor, although the inventor will hold the patent personally. How this works depends upon state law. If you become an inventor, you will need to learn the state laws relevant to your place of employment. Remember: Legal research in an unfamiliar area can be difficult to conduct, so the best place to get advice is from an attorney's office.

What happens if someone infringes upon a patent during its life? Violators are punishable in the federal court system. Penalties can be severe and may include damages for loss of sales of the product and even of related products. Even if the inventor can show no monetary loss, the court may award damages based upon the value of the infringement to the violator. This may include profits earned by the infringer's using the patented product as well as reduced development costs incurred. The inventor may be able to recover legal fees as well. In addition, the infringer may be ordered to cease production and sales of the product. If it can be shown that the violation was willful (e.g., counterfeiting), the court may award enhanced damages (e.g., treble damages).

Clearly, the patent world is fraught with legal issues. If you find yourself in that world, you must be keenly aware of the wisdom of engaging a competent, experienced patent attorney.

COPYRIGHTS

Copyrights are a form of federal protection for original works of authorship when they become fixed in a tangible form of expression. Works that are subject to copyright protection include

- literary works (novels, newspaper or magazine articles, textbooks)
- musical works, including lyrics
- dramatic works, including musical accompaniment
- pantomimes and choreographic works
- Art works (pictorial, graphic, and sculptural works)
- motion pictures and other audiovisual works
- sound recordings
- architectural works
- computer programs

Copyright protection applies to the form of material expression of ideas but not to the ideas themselves. Therefore, you (or I as co-author of this book) are free to conduct research on any given topic; read literature; or textbooks or newspaper, magazine, or Internet articles on a topic; and then to express what we have learned in our own words and in our own way. We can create expressions of similar ideas, but we cannot legally reproduce the work or create derivative works based upon the original work, nor can we distribute copies of the original work to the public. As an example of the power of copyright protection, usually when you hear a popular song on the radio or TV or in a club or community theatre production, the station or the theater is paying a fee (royalty) to the holder of the copyright.

Copyright protection can also apply to various business documents such as brochures, training manuals, sales manuals, advertising copy, and more. Although some computer software is categorized as being in the public domain, most computer software is copyright protected and can be legally installed only after purchase of a license, usually attached to the software when it is purchased. Unless a license allows the software to be installed in multiple computers, you need to purchase a license for each computer into which you wish to install the software. Some very serious penalties have been imposed upon engineering firms that have violated copyright laws by installing software on multiple computers without purchasing additional licenses.

Copyright protection is by far the simplest and easiest form of intellectual property protection to obtain. It is basically automatic. A copyright comes into existence at the moment an original work takes a tangible form, although protection is limited. No formal application is required; although an optional registration process can be followed, resulting in stronger protection. You can apply for a copyright for a work from the U.S. Copyright Office. Up until 1989, a formal registration was required. In 1989, the United States signed on to the Berne Convention for the Protection of Literary and Artistic Works, usually known as the Berne Convention, an international agreement about copyright first adopted in Berne, Switzerland, in 1886. Thereafter, copyright protection became automatic, with no registration required.

Placing a proper statutory notice of copyright on all copies of a work distributed to the public is legally necessary. Acceptable forms of notice of copyright contain the copyright symbol (©). Two common forms of notice of copyright are:

- ©, year of first publication, name of copyright owner (©, 2010, Anyone Publishing Company); or
- ©, year of publication, author (©, 2010, John Smith).

Notice of copyright should be put in a prominent place on each copy of the work. This is not a legal requirement for works first published after March 1, 1989, but it is good practice. The value of notifying users of the work is that it may deter them from making unauthorized use of the work, and it may become an important piece of evidence should a copyright infrigement action be taken to a court of law. Sometimes an infringing party can convince the court that he was unaware of the copyright, resulting in a reduced penalty.

In civil actions brought for copyright infringement, the infringing party can experience severe consequences. The court can order forfeiture or destruction of the infringing materials and can even go so far as to order that the equipment used in producing the materials be forfeited or destroyed. Significant monetary damages can be awarded to the plaintiff, the owner of the copyright. In addition to civil lawsuits over alleged copyright infringement, there can be criminal prosecution for alleged violation of copyright law.

You may be interested to know how long copyright protection lasts. The duration of copyright protection has varied throughout history, but since January 1, 1978, in the United States the duration has been set at the life of the author plus 70 years, although there may be some variation based upon the nature of authorship (joint works, anonymous works, works made for hire, etc.). After expiration of the copyright protection on any work, it can be used freely by anyone. It is then considered to be in the public domain. The public domain is that body of knowledge considered to be part of our cultural and intellectual heritage. There is no restriction for use of works that are in the public domain.

Some works cannot be copyrighted by law. They include U.S. government publications, statutes, judicial opinions, administrative rulings, and the like. The same principle holds for similar state and local works.

Before we leave this section, we need to talk a bit about copyright protection as it applies to engineers. In 1990, Congress passed the Architectural Works Copyright Protection Act. The protection lasts for 95 years from the date of publication of the work, or 120 years from the date of the creation of unpublished work. It applies to all buildings constructed after

December 1, 1990, although the courts have held that the act covers buildings under construction at the time it became law.

Since the enactment of the law, courts have interpreted it to include other elements of the design process as eligible for copyright protection, including technical drawings, site plans, and maps. Copyright owners have exclusive rights to reproduce, distribute, prepare derivative forms, and to display copyrighted work. Interpreted literally, no one has the right to reproduce architectural works or to rely upon them as the basis for additional design in the future (unless agreed upon by contract).

For engineers, the unauthorized reuse of a site plan for additional design represents a liability risk. To the extent possible, contract language should prohibit unauthorized reuse of the work. It is advisable to provide additional protection by displaying a notice of copyright on the plans. To strengthen the protection, the engineer should register the plans with the U.S. Copyright Office. Timing is important, as the legal protection gained by registration can be diminished if the filing of the copyright registration is not at the proper time. Filing takes time and costs money, and many engineers don't do it.

The topic of copyright protection for architectural works is presented here to give you an awareness of its existence. It is a complicated matter. I have had no personal experience with it in my own practice, and we are not educated in all the nuances of the law and of court cases. I want to emphasize that this book cannot be used as a substitute for competent legal advice. Our goal is to give you a broad background, familiarizing you with many aspects of practice that may arise during your career. In the matter of copyright protection, we suggest that you attend any seminars at which it will be discussed and that you conduct independent research when and if the occasion arises that you want or need to seek copyright protection.

TRADEMARKS

Trademarks are words, names, symbols (logos), slogans, or designs—or more than one of those in combination—that identify a company's products or services, distinguishing them from the products or services sold by others. In some circumstances, a trademark used exclusively by a service organization may be known as a service mark. Trademarks can also be used in various ways for people to show membership in an organization.

The term *trademark* is generally equivalent to the terms *brand, mark,* or *logo.* All those terms describe a distinguishing graphic symbol, word or words, or design used alone or in combination to identify a business, a product, or service. You are familiar with the famous Jaguar hood ornament, the Apple Computer Company symbol, the America Online Instant Message runner, and so many more. They are a natural part of your everyday living. They, like all trademarks, immediately bring the products to mind, even if seen out of context.

A trademark may be identified by the symbol ™ to indicate that the owner claims exclusive rights to the trademark but that it is not registered. If the trademark is legally registered, it will be shown with the symbol ®.

Trademarks are treated in law as a type of property. Ownership of a trademark can be created only by actual use, but registration is recommended. Registration of a trademark with the USPTO gives the owner exclusive rights to its use. It also gives the owner the right to prevent unauthorized use of the trademark.

To register a trademark, the owner must demonstrate that it has a distinctive character. If it does not appear to the USPTO examiner that it has a distinctive character, the owner must demonstrate that the public has formed an association of the mark with a product or service. That association is termed "acquired distinctiveness."

Trademark registration, unlike copyright registration, does not provide a long-term period of protection. It is possible to lose the protection. For instance, if the owner fails to use the trademark over a period of time, say three to five years, the trademark may be removed from the register. It can then be considered in the public domain and can be used by anyone. It can also be reregistered by anyone, including the owner.

As long as the trademark remains registered, the owner has a very good chance of preventing its unauthorized use. Unregistered trademarks are more difficult to control.

Before attempting to register a trademark, the owner should conduct a search to be sure the design is distinctive, meaning there is no likelihood of confusion with another trademark. Other trademarks may exist that are closely patterned after the desired design. This can apply to the choice of business name as well. Some attorneys are experts in the area of trademark law. They can make searches through a variety of record systems, including databases. When the search has reached a reasonable level of thoroughness, the attorney can review the findings and give an opinion as to the probability that the trademark is or is not free for use. If the opinion is favorable, then it is time to register.

Trademark owners can enforce their rights by filing a complaint against alleged infringers in state or federal court. The judge assigned to the matter can order the alleged infringer to cease use of the trademark, even before the trial commences. The judge can also pull allegedly counterfeit products from the marketplace and cause the infringing products to be seized by law enforcement officials. If the trademark owner prevails in court, the infringer's products can be destroyed, and monetary damages can be awarded to the owner. If the infringement is found to be willful, the court can award treble damages. In addition to a civil trial, the infringer can be charged and tried as a criminal. If convicted, a fine of up to $2 million can be imposed, and the infringer can be sent to prison.

Needless to say, the protection of trademark rights is zealously pursued. Those who choose to make use of a protected trademark illegally are exposing themselves to very harsh penalties.

TRADE SECRETS

A trade secret is any practice, process, formula, design, instrument, pattern, or compilation of data, the use of which results in a competitive advantage for a business over its professional or industrial competitors. All companies in the world may have trade secrets. The international conventions and laws regarding protection of trade secrets are beyond the scope of this book. In this chapter, we focus on the conventions and laws of the United States.

Companies desire to protect their trade secrets and, in some cases, will go to great lengths to do so. Perhaps the most prominently recognized trade secret is the formula for Coca-Cola. Reportedly, it is securely enclosed in a bank vault that cannot be opened without a corporate resolution passed by the board of directors of the Coca-Cola Company. Only two Coca-Cola employees, whose identities are never made known to the public, are entrusted with the formula at any given time. Those employees are restricted from traveling together on the same airplane. Such measures are extreme and far exceed what an average American corporation will do to protect its trade secrets. But protect them they must.

A trade secret must be managed in a manner that can be reasonably relied upon to prevent competitors, or any member of the public, from learning it. If a company seeks damages in court because of an alleged theft of a trade secret, it must be able to demonstrate that adequate measures were taken to protect the trade secret.

Let's look at some examples of types of trade secrets that a company would want to protect:

- A formula for a product, such as a beverage (Coca-Cola), a glue or bonding agent, a medication, a coating, and so on
- A unique or innovative sales or marketing program
- A manufacturing process
- Lists of clients, customers, or suppliers
- Marketing survey results and analysis
- Innovative methods for product distribution
- Future models of products

Trade secrets, unlike the other forms of intellectual property, cannot be registered. Trade secrets can be protected for unrestricted periods of time. For that reason, a company might choose to keep an invention in the trade secret category rather than patent it, because patents eventually expire. When a patent expires, the patented item enters the public domain, but a trade secret, if it can be kept secure, will last forever.

Certain criteria must be met for a company to establish trade secret status for its information:

- The information must not be generally known or available to those who conventionally deal with that type of information.
- The information's secrecy must lend commercial value.
- The holder of the trade secret must demonstrate that reasonable steps have been implemented to maintain secrecy.

Companies can take some defensive measures to protect their trade secrets. They can make sure that the documents or devices that comprise trade secrets are carefully and securely protected, particularly during periods outside of regular business hours. Also, trade secrets should only be shared within the company on a need-to-know basis.

Beyond those obvious steps, companies can provide the best trademark secret protection through implementing nondisclosure agreements with employees and, in some cases, with vendors, customers, service providers, or entities with which they are doing collaborative work that, because of their relationship with the company, may have access to some of its trade secrets.

As a young engineer, you may be asked to sign a nondisclosure agreement, sometimes known as a confidentiality agreement, by an employer that is concerned about trade secrets. A nondisclosure agreement is a contract that binds both parties to its terms and conditions. The purpose of a nondisclosure agreement is to protect the company from someone's revealing trade secrets during and, sometimes, after employment. Once you sign a nondisclosure agreement, you are legally bound to its terms and conditions. An oral agreement is legal, but, as indicated in Chapter 33 on contracts, it is a very bad idea.

A nondisclosure agreement should contain five basic elements:

1. *Specification of what constitutes confidential information for purposes of the agreement.* A list of categories of information should be given along with a list of specific areas within each category.

2. *Specification of what is excluded from confidential information.* For instance, if you brought an idea or invention that was conceived or created on your own before coming to the company, you might be wiling to share it with the company, but you might want it excluded form the agreement.

3. *Obligations to the receiver (in this case you).* These should be clearly stated.

4. *Duration of time that you must maintain secrecy.* The company will likely want it to be perpetual, while you may wish it to be shortened to some reasonable time.

5. *Miscellaneous provisions.* These may include alternate dispute resolution forums (arbitration, mediation), which state laws will govern any dispute, and whether attorney fees be paid by the losing party to the prevailing party.

If you are presented with a nondisclosure agreement, be sure to read it carefully and, if possible, have it reviewed by an attorney of your choice. Be ever so careful in attorney selection. Your family lawyer may not be familiar with trade secret law. It is advisable to seek the opinion of a patent attorney with experience in that area.

Does all of this discussion of contracts really matter? You bet it does. There are trademark protection laws in over 40 states. Most are based upon the federal Uniform Trade Secrets Act. These laws permit a trade secret owner to file a civil suit against the alleged violator. Some states have criminal laws covering theft of trade secrets as well. You can see which states have trade secret laws by going to the Cornell Law School's Web site at *www.law.cornell.edu/uniform/vol7.html*.

In addition to state laws, the Economic Espionage Act was signed into law on October 11, 1996. The introductory sections are reproduced below. Section 1831 deals with violations of the act that benefit foreign nations, whereas section 1832 focuses on both foreign and interstate commerce.

UNITED STATES CODE

Title 18 – Crimes and Criminal Procedure

Part II – Criminal Procedure

Chapter 90 – PROTECTION OF TRADE SECRETS

Cite as the "Economic Espionage Act of 1996"

Sec.

1831. Economic espionage.

1832. Theft of trade secrets.

1833. Exceptions to prohibitions.

1834. Criminal forfeiture.

1835. Orders to preserve confidentiality.

1836. Civil proceedings to enjoin violations.

1837. Conduct outside the United States.

1838. Construction with other laws.

1839. Definitions.

§ 1831. Economic espionage

(a) In General.— Whoever, intending or knowing that the offense will benefit any foreign government, foreign instrumentality, or foreign agent, knowingly—

(1) steals, or without authorization appropriates, takes, carries away, or conceals, or by fraud, artifice, or deception obtains a trade secret:

(2) without authorization copies, duplicates, sketches, draws, photographs, downloads, uploads, alters, destroys, photocopies, replicates, transmits, delivers, sends, mails, communicates, or conveys a trade secret:

(3) receives, buys, or possesses a trade secret, knowing the same to have been stolen or appropriated, obtained, or converted without authorization:

(4) attempts to commit any offense described in any of paragraphs (1) through (3); or

(5) conspires with one or more other persons to commit any offense described in any of paragraphs (1) through (4), and one or more of such persons do any act to effect the object of conspiracy shall, except as provided in subsection (b), be fined not more than $500,000 or imprisoned not more than 15 years, or both.

(b) ORGANIZATIONS.— Any organization that commits any offense described in subsection (a) shall be fined not more than $10,000,000.

§ 1832. Theft of trade secrets

(a) Whoever, with intent to convert a trade secret, that is related to or included in a product that is produced for or placed in interstate or foreign commerce, to the economic benefit of anyone other than the owner thereof, and intending or knowing that the offense will injure any owner of that trade secret, knowingly—

(1) steals, or without authorization appropriates, takes, carries away, or conceals, or by fraud, artifice, or deception obtains such information;

(2) without authorization copies, duplicates, sketches, draws, photographs, downloads, uploads, alters, destroys, photocopies, replicates, transmits, delivers, sends, mails, communicates, or conveys such information;

(3) receives, buys, or possesses such information, knowing the same to have been stolen or appropriated, obtained, or converted without authorization;

(4) attempts to commit any offense described in paragraphs (1) through (3); or

(5) conspires with one or more other persons to commit any offense described in paragraphs (1) through (3), and one or more of such persons do any act to effect the object of the conspiracy, shall, except as provided in subsection (b), be fined under this title or imprisoned not more than ten years, or both.

(b) Any organization that commits any offense described in subsection (a) shall be fined not more than $5,000,000.

During your career, you will continuously add to your personal body of knowledge. As repeatedly said, this book is not intended as a substitute for competent legal advice. Our intention is to sensitize you to some of the tricky areas that you may encounter during your career.

32

QUALIFICATION-BASED SELECTION

Qualification-based selection, or QBS, is a method used by many owners of projects to procure engineering design services. It is an alternative method to soliciting bids for engineering work on design projects. It is, as you will see, the most prudent and fair method for procurement of design services.

WHY QBS IS PREFERABLE

Let's examine the underlying principles that support our preference for QBS over bidding.

When most owners decide they want something to be designed and constructed, they have little detail available as to what the final product will be. It is widely recognized, however, that the cost of engineering design is a small percentage of the construction costs required to bring most projects to completion. If one were to consider overall life cycle costs, including operation, maintenance, and financing, the engineering costs become even less significant. Yet at the inception stage, the owner often has little more than a project scope document with little detail.

The ability of an engineering firm to accurately forecast the investment in man hours, technology, and outside consultant support is limited. In fact, it is virtually impossible. The consequences of procuring design services through competitive bidding often results in the best firms not submitting bids, and those that do are risking winning a project that can result in significant cost overruns. Cost overruns put a significant strain on the firm, placing it in an untenable position that may require reduction of costs in whatever way possible. Such circumstances can lead to reduction of quality in the finished design and increased professional liability exposure.

CODES OF ETHICS AND THE LAW

There was a time when various engineering societies considered bidding an unethical practice and said so in their codes of ethics. In 1974, however, the U.S. District Court for the District of Columbia ruled as follows:

By order of the United States District Court for the District of Columbia: Former Section 11c of the NSPE Code of Ethics prohibiting competitive bidding, and all policy statements, opinions, rulings, or other guidelines interpreting its scope, have been rescinded as unlawfully interfering with the legal rights of engineers, protected under antitrust laws, to provide price information to prospective clients; accordingly, nothing contained in the NSPE Code of Ethics, policy statements, opinions, rulings, or other guidelines prohibits the submission of price quotations or competitive bids for engineering services at any time or in any amount.

The case went on to appeal before the U.S. Supreme Court, which affirmed the district court's decision.

The NSPE Code of Ethics was, of course, revised to reflect the finding of the courts. It should be noted, however, that in its decision to affirm the district court's decision, the Supreme Court noted that engineers and engineering firms may individually refuse to bid for engineering services; that clients are not required to bid for engineering services; and that federal, state, and local laws governing procedures to procure engineering services are not affected and may remain in full force and effect. This information is included as a footnote to the NSPE Code of Ethics as published.

The 1972 Brooks Act (P.L. 92–582) requires that the federal government shall use QBS for procuring architectural and engineering services on federal projects. The Federal Acquisition Regulations (FAR) further detail the requirement for QBS in procurement of architectural and engineering services. Forty-seven states and many local governments have adopted laws and regulations requiring the use of QBS on their projects. Let's examine what QBS is all about.

HOW QBS WORKS

An owner who uses QBS on very large projects will usually set up a selection committee staffed with highly experienced, savvy people. They do not necessarily all have to be employees of the owner. There is no single model, but the committee usually includes the owner's construction manager, the person who will manage operations at the intended facility, a professional engineer, the owner's primary contact person with design professionals, and, for building projects, an architect. Anyone with specialized knowledge of the project under consideration may be added to the mix. For smaller projects, the selection committee may not be as formal, and the owner may choose to fulfill the functions of the selection committee.

The selection committee will eventually be called upon to judge the qualification of the engineering firms who express interest in the project. The committee, to be as objective as possible, will establish a set of criteria by which applicants will be judged. The characteristics the committee will definitely want to include are evaluation of the firm's experience with similar projects, staffing levels, and the staff's professional credentials; the availability of firm resources during the anticipated project term; and any other issues of importance to the project.

When the committee is ready to act, the owner will distribute a request for qualifications (RFQ) to those firms it considers eligible. Perhaps an advertisement for interest may be placed in local or regional newspapers or other media. Interested firms will receive documents indicating the areas of interest established by the committee.

Before issuing RFQs, the owner must prepare a project scope document. It should be as informative as possible. It should outline the type and size of the intended facility; the preliminary budget and scheduled constraints; site-specific information regarding known environmental issues, soils, or survey data; and anything else the owner can furnish to help interested firms grasp the project's scope. If the owner is not professionally staffed to produce a project scope document, it should engage an independent professional to prepare one.

The scope document and RFQ are distributed as described above. Interested firms respond to the RFQ by providing the data specified. A cutoff date for acceptance will be set. At this stage, no fee quotations will be solicited.

When firms have submitted their statements of qualifications, the selection committee evaluates them against the criteria it has set. It may also contact references, if they were required. Normally the selection committee chooses a minimum of three firms as the most qualified, although it may select more than three. Then the committee sends the top firms requests for proposals (RFPs). The RFP solicits additional information from the firms about key personnel to be used on the project, the office location where the work will be performed, the firm's financial strength, an outline plan for conducting the work, and other information to use in ranking the proposals. No fee proposals will be solicited at this time either.

When the selection committee has received all of the proposals, it sets about ranking them based upon detailed review and analysis. Once that process is complete, the selection committee turns over its work product to the owner and ordinarily disbands. The owner then meets with the top-ranked firm to discuss the project in detail. The owner and engineer collaborate to work toward a refined scope of services, each bringing to the table perspectives gained through their years of experience. Through that process, both owner and engineer become familiar with the thoughts of the other and adjust their perceptions to accommodate mutual interests.

At the conclusion of the refinement of scope phase, the parties begin fee negotiations. The engineer defends the fee proposal from a designer's perspective, while the owner is likely to challenge it based upon experience with similar projects in the past. If a meeting of the minds occurs, the parties move into the contract stage, described in Chapter 33. If they cannot come to a mutually agreeable position, the owner terminates the negotiation and moves on to the second-ranked firm. The parties do essentially the same things as during the first interview and negotiation. If the parties can't come to agreement, the owner moves on to the next firm. This process continues until an agreement is reached or until no firms are left with which to negotiate.

QBS allows the engineer and the owner to jointly examine the project in detail. Out of that process comes a very highly refined scope of work. Both parties have the benefit of hearing each other's detailed thinking. In a bidding process, however, that exchange of ideas is not possible. There may be some interaction between the bidders and owner through requests for information, site inspection, and possibly a prebid meeting between the owner and all the bidders. But that level of communication cannot measure up to the collaborative interaction between the owner and individual firms, as in the QBS process.

Some prominent national associations and agencies support QBS. They include the following:

- American Bar Association
- American Council of Engineering Companies
- American Public Works Association
- American Institute of Architects
- American Society of Civil Engineers
- Federal Highway Administration
- National Society of Professional Engineers (see policy in the appendix)
- United States Army Corps of Engineers

QBS is clearly the most effective way for owners to procure engineering services. The private sector displays some resistance to the process, especially from unsophisticated owners who view engineering services as a commodity. The major engineering societies listed above make continuous efforts to educate owners about the value of QBS and to promote its use over bidding. NSPE and ACEC jointly sponsor QBS awards annually. One is given to the public agency operating under QBS legislation that exemplifies the value of the process, and one is given to an owner who voluntarily uses QBS in the absence of any legal requirement.

It is important for you to understand the value of QBS procurement. You may not be in a position to influence the manner in which your employer makes proposals for design projects, but someday you may be in a position to add some meaningful dialog within your firm, at a client's office, or at a society meeting. That's what this book is all about.

33

CONTRACTS

GENERAL INTRODUCTION

At this early stage in your career, you may believe that contracting for engineering services is irrelevant to your job and that you don't need to spend any time on it. However, although contracting may not apply to you now, it could be just a few years away, so you need to be prepared for that stage in your career.

Engineers must have a basic understanding of contracts, whether they are in private practice as design professionals or as constructors in government or in industry. Engineers in education should also have an interest in contracts if they are engaged as consultants for research, or if they enter into research grant agreements.

A contract, according to *Black's Law Dictionary*, is

An agreement between two or more persons which creates an obligation to do or not to do a particular thing. Its essentials are competent parties, subject matter, a legal consideration, mutuality of agreement, and mutuality of obligations.

A consideration is

An inducement to a contract, the cause, motive, price, or impelling influence which induces a contracting party to enter into a contract.

A legal consideration is one recognized or permitted by the law. Therefore, a contract that would obligate one party to perform a service in exchange for an illegal act, say to harm a third party, or to do some legally forbidden act, would not be acceptable. An engineer cannot legally enter into a contract to prepare plans in a manner inconsistent with law. For instance, an engineer cannot legally agree to draw plans for a development in a regulated wetlands area without applying for the necessary environmental permits.

This chapter will discuss some issues with respect to engineering contracts. The variety of contract types is voluminous. A list of just some contract types (also known as agreements) that may become part of an engineer's professional life includes

- confidentiality/nondisclosure agreements
- construction contracts
- consulting contracts (owner/engineer)
- employee stock option agreements
- employment agreements
- engineer/subconsultant contracts
- incentive plan agreements
- incorporation certificates
- indemnification agreements
- joint venture agreements
- lease agreements
- limited liability company agreements
- management agreements
- merger agreements
- noncompetition agreements
- partnership agreements
- research and development agreements
- severance plan agreements

The list is not exhaustive as to the total spectrum of contract or agreement types. Most commercial activity is controlled by a variety of contract types. For purposes of this chapter, we will concentrate on some of the contract types that will be most prominent in a young engineer's developing career, depending on sector of practice. An understanding of contracts is a wonderful asset to any engineer who is interested in climbing the career ladder, whether on the technical path or the business management path.

For those engineers who are engaged in the industrial side of engineering, seldom is there a written contract between their firm and their customers, the general public. Contracts may, however, be required between commercial or industrial companies and outside consultants or suppliers enlisted to assist with product development. Many such design firms in the private sector offer engineering services to clients who have an idea about a product but have some reason for wanting the design, testing, and—sometimes—manufacturing of the product to happen outside of their organization. In those circumstances, the product development firm usually makes a proposal and enters into a written contract with its client.

For those engineers who are engaged in private practice, their view of contracts will normally be from the consultant's perspective. Their contracts are usually with clients or subconsultants. Such contracts generally define the scope of the project to be designed, the duties and obligations of the engineer and the owner, compensation and payment schedules, and more. A more detailed list follows later in this chapter.

For those who are engaged in government engineering, their view of contracts will normally be from the owner's side of a project. By that, I mean that your government employer (municipality, county, state, federal agency, etc.) may be engaging an outside design professional, manufacturer, vendor, material supplier, or construction contractor to provide goods or services to your governmental unit. Your organization is the purchaser rather than the supplier.

For those engineers who are engaged in the construction industry, their view will be from the perspective of the builder. Their contracts are usually with the owners of the projects they will construct and with subcontractors, materials suppliers, equipment rental companies, and—sometimes—surveyors or engineers engaged for project support.

What should a well-constructed contract do for each party entering the agreement?

The first and foremost element of a good contract is that it is a formal written document. While every course ever taught in contracts says that oral contracts are just as valid as written ones, the weakness of oral contracts is pretty obvious. Verbal agreements can be open to interpretation, and there is no way to prove what was promised by either party.

If all goes as planned and as agreed by the parties, the oral contract will have served its purpose. Sometimes, however, one party to an oral contract feels that the other has failed to meet its obligations under the contract, potentially resulting in litigation. Then the parties will be put to the test of proving their interpretation of the unwritten agreement to a court. Obviously, the issue becomes one of credibility, and it can get expensive if lawyers and witnesses get involved. Having a written document will help eliminate the he-said/she-said of any agreement. Written contracts can also result in disputes, but they offer concrete evidence essential to any proceedings.

CONTRACTS PRIMER FOR ENGINEERS

For purposes of this discussion, the parties to a primary contract will be called Owner and Engineer. *Owner* will mean that person or entity that initiates a design or construction project and will provide financial resources to fund it. *Engineer* will mean the person or legally sanctioned firm that will perform specified services, including, but not limited to, predesign studies and planning, preparation of program documents, plans and specifications, calculations, reports, environmental permit applications, construction or manufacturing support services, and, sometimes, facility start-up.

Occasionally, multiple design entities work on a project. They might include an engineer, an architect, a land surveyor, a geotechnical consultant, an environmental specialist, a structural engineer, and more. In those circumstances, the group would be called a project team. Often, one of the members of the design team will be designated by Owner as Prime Professional and will have a contract directly with Owner, while the others will be subconsultants and have contracts with the Prime Professional. All of the other members of the design team may also have direct contracts with Owner. Such arrangements are described as multiprime contracts.

Whatever the arrangement, each party to a contract must recognize certain important principles. In the introduction to this chapter, we touched on some of them. Now we'll get a bit more detailed.

TYPES OF CONTRACTS

We will look at five types of contracts. All can be used, although some are preferable for engineering projects. The five types of contracts are:

1. Oral contracts
2. Letter form agreements (proposals)
3. Purchase orders
4. Custom contracts
5. Standardized form contracts

Oral contracts. As stated earlier, while oral contracts are acceptable for most transactions, they are not advised. When things go well and no disputes arise, oral contracts serve the purpose. When things do not go well, however, disputes can only be resolved through litigation. The process will be difficult, as little clear evidence of the intent of the parties will exist. Needless to say, oral agreements should be avoided to the maximum possible extent.

Letter form agreements (proposals). Frequently, Engineer may be asked to respond to a request for proposal (RFP) issued by Owner. Generally, Engineer will draft a proposal in letter form to Owner. The proposal will define the project and set forth the scope of service Engineer intends to perform, along with other pertinent information including but not limited to

- obligations of Owner
- services specifically excluded
- time allotted for provision of services
- compensation terms for proposed scope of service
- provisions for authorization of extra work and compensation therefrom
- specific exclusions, such as special certifications, guarantees, and warranties
- specialized service beyond Engineer's normal capabilities, etc.
- ownership of documents
- dispute resolution procedures
- provisions for termination of contract

In letter form proposals, some engineers will attach a list of specific terms and conditions, sometimes termed *boilerplate,* "a term describing uniform language used normally in legal documents that has a definite, unvarying meaning in the same context that denotes that the words have not been individually fashioned to address the legal issue presented" ("Legal Encyclopedia" at Answers.com). Usually, such terms and conditions will have been developed by the engineering firm, with advice of counsel over a period of time, and will reflect lessons learned by experience. A firm's standard terms and conditions are generally maintained and updated as required by external circumstances, such as modifications to or creation of new laws, regulations, and standards and technological advances.

If Owner accepts Engineer's proposal, they preferably will draft a formal contract incorporating the final form of proposal, even if only by reference to it. Often, however, Owner simply signs and accepts the letter form of proposal. Any amendments to the proposal can be written onto the form and initialed by both parties. Whenever this technique is used, the letter form proposal should be as detailed and comprehensive as Engineer can produce.

Purchase orders. Sometimes, Owner will respond to a letter form proposal by orally authorizing work and then issuing a purchase order to Engineer. Some organizations need to have a purchase order number for their accounting departments to process invoices for payment. This is a perfectly reasonable thing from a payment perspective, but the standard purchase order is seldom acceptable for entering into contract with an engineer.

Use of a purchase order can present difficulty for Engineer, because purchase orders are standard documents prepared for the purchase of goods and services other than engineering (such as office supplies, equipment, plumbing repairs, construction, maintenance, etc.) They usually have stringent requirements that are not applicable to engineering services. Commonly, they will have provisions for guarantees and warranties as well as broad indemnification clauses in favor of Owner. Such language is unacceptable in engineering contracts and should be meticulously stricken from the purchase order. Such changes will, of course, require acceptance by Owner.

If Owner insists upon using a purchase order, it is preferable to draw a contract that will satisfy both parties and eliminate the onerous provision of the purchase order in the contract. If Owner resists, Engineer has a business decision to make. It is important to be aware that some of the provisions in purchase orders place uninsurable requirements on Engineer. For instance,

Engineer's professional liability insurance will exclude coverage for claims made for breach of warranty or guarantee.

Purchase orders require special vigilance and should be used only after detailed review and modification to bring them into compliance with the commonly recognized terms in engineering contracts.

Custom contracts. Custom contracts are agreements that are crafted by one of the parties and negotiated into final form by both parties. Both parties must keep in mind that they are creating a document that will set the guidelines for what should be a mutually beneficial project. Risk allocation, a central theme of any contract, should be done fairly. Generally speaking, a party should be responsible for only that which it can control. For instance, Owner will control the construction site, so Engineer should not be asked to indemnify Owner from any and all hazards that may arise at the construction site.

A custom contract should cover a broad variety of issues. It must be emphasized that, whereas Engineer needs to be involved in the negotiations of the contract content, advice of counsel is always prudent. A professional liability insurance company may also provide assistance in contract preparation. Some of the areas of concern in a custom contract are the following:

- Scope and nature of project
- Scope of Engineer's service
- Responsibilities of Owner
- Owner-provided documents, surveys, studies
- Payment schedule (compensation) for basic services
- Authorization and payment for extra services
- Reimbursement of expenses
- Change of scope provisions (financial and time extension)
- Time for provision of services
- Names of authorized persons for Owner and Engineer
- Termination provisions
- Engineer's role during construction
- Assignment of contract
- Indemnification provisions
- Limitation of liability
- Methods of dispute resolution

- Standard of care to be expected of Engineer (Important: The contract should not offer perfect service or any form of warranty or guarantee.)
- Suspension of service for failure to pay professional fees

The list of important issues could be expanded, but this should be sufficient to give you an idea of what to look for.

Standardized form contracts. By far the best option open to Engineer is to use standard form contracts prepared by the Engineers Joint Contract Document Committee (EJCDC). The EJCDC is staffed by highly skilled and experienced volunteers and staff members from the National Society of Professional Engineers (NSPE), the American Council of Engineering Companies (ACEC), the American Society of Civil Engineers (ASCE), and the Associated General Contractors of America (AGC). The committee produces a family of fair and balanced contract documents, all of which are coordinated and consistent. To perform such a mammoth task, the committee meets several times each year to continuously review and upgrade the documents with input from organizations throughout the construction industry.

The EJCDC documents are widely recognized by owners, engineers, and contractors as well as the court system. The entire family of documents is coordinated and consistent with respect to the duties of the parties. There are documents in the following categories:

- Construction
- Owner/engineer
- Engineer/subconsultant
- Funding agency
- Design/build
- Procurement
- Environmental remediation

To examine the content of the individual sets and contract documents, visit *www.nspe.org/ejcdc/home.asp*.

It should be noted here that the American Institute of Architects (AIA) has also developed a set of standard contract documents widely used by that profession. On occasion, owners may attempt to use these documents in the engagement of an engineer, and sometimes an architect will offer a subconsultant role to an engineer using the AIA documents. Engineers should recognize the existence of both sets of standard contract documents. Any attorney practicing in service to the two design professions should be knowledgeable about the two sets, as should an insurance broker specializing in PLI. Be aware of the AIA documents. While similar to the EJCDC documents, they are not identical.

For those engineers engaged in product development design, some of the information regarding standard of care may not be applicable. Products are held to a different legal liability standard (strict liability) than that applied to architects or engineers who provide services to the public. Dealing with that subject is a bit beyond the scope of this book.

CONCLUSION

Like so much that awaits you in your engineering career, contracts may take up more of your time and knowledge when you become a manager. Even as a student or apprentice engineer, however, you should grasp the fact that contracts are an important part of commercial life and that contracts for engineering services can be very complex. Remember that contracts are a means of allocating risk as well as the road maps for the performance of an engineering firm during a project.

While many firms keep contracts under lock and key, keeping a copy of the pertinent sections of a contract in a project file is advisable, so that the project manager and the staff have the benefit of seeing what was promised to the client. If your firm does so, take the time to read its contracts, and if you think the firm is deviating from its responsibilities under a contract, bring your concerns to the attention of your supervisor. More often than not, your respectfully expressed concern will show that you have an interest in the project and in the firm beyond that of someone who is working only for a paycheck. You will begin to show that you are a true professional and that you care about the firm.

34

DISPUTE RESOLUTION

As a young engineer, I thought that as long as I was careful with my work, there would be little chance of error and therefore little chance for anyone to claim damages from my work. That was a naive dream. Idealism is a wonderful attitude, and for every engineer to be as thorough and accurate as the situation demands is a worthy goal, but all the idealism in the world will not guarantee that a claim won't arise from your work.

Unfortunately, all persons in the engineering profession will probably see disputes arise during their career—particularly in the areas of private practice consulting, construction, and government. In Chapter 28 on risk management, we looked at ways to reduce the likelihood of disputes, but inevitably they will arise. When that happens, you need to understand that multiple approaches are available to resolve them and, as a young engineer, you can play a role in the process.

Dispute resolution strategies range from early action in the form of discussions at the first sign of a problem, all the way up to litigation (a trial in a court of law). Dispute resolution strategies can be classified into two overarching categories: adjudicative and nonadjudicative.

Adjudicative dispute resolution refers to a process in which the parties present their case to a third authoritative party, who will render a decision that is binding upon the parties. The two forms of adjudicative dispute resolution are litigation and arbitration.

Nonadjudicative dispute resolution techniques include informal discussion, negotiation, use of predetermined arbiters, and mediation. The cost of adjudicative dispute resolution is usually quite high in terms of time, money, and emotional stress. The costs of early, nonadjudicative dispute resolution are always far less than those of adjudicative processes.

We will first explore the various strategies, working our way up from the least costly to the most costly. We will begin looking at the nonadjudicative procedures and what your role might be in this process as a young engineer.

In Chapter 28, we explored some of the causes for disputes and laid out some red flags that can indicate potential problems. You may be in a position to sense an impending problem even before your project manager or the principals of the firm. As a responsible member of your organization, you should point out the issue to your supervisor or project manager.

If you have ever canoed white water, you will recognize the term *reading the river*. A small ripple in the flow of a river can indicate that a large rock is looming just below the surface. Usually the person in the bow is charged with seeing that and signaling to the person in the stern, who has most of the control over the course of the canoe. When a small ripple occurs in a project, you may be the bow person, and you may need to get the information to someone who has more control within the firm. You may be able to signal a looming crisis. Train yourself to focus on such potentials—to "read the river." That skill will go a long way toward enhancing your status in the organization as well as toward helping the firm to safely navigate the shoals.

To illustrate, let's create one possible scenario. Assume you are assigned to a project team under the supervision of a seasoned project manager. A client has engaged your firm to prepare plans and specifications for site development in support of a building expansion. You are asked to sit in on periodic project meetings with your project manager and with the client's project team. One of the client's people in regular attendance is the plant facilities manager. The meetings are all cordial and positive in tone, with no significant problems arising. Your project manager assigns you to have periodic telephone conferences with the client's facilities manager. During one of the calls, he expresses mild dissatisfaction with progress of the project and makes some vaguely negative comments about your firm's invoicing process. Because none of this ever came up at the project meetings, you are puzzled.

You may have just identified a ripple. Perhaps the facilities manager has picked up some negative vibes at meetings within his own organization. You are aware that you are not in a position to address his comments from your position as a junior engineer, nor should you try to do so. You can, however, having read a ripple in the river, communicate it to someone who does have some degree of control. You should report the conversation to your project manager as factually as possible. What you heard might be simply balder-dash, but you can't be sure. By reporting it to a manager, you may be giving the firm a well-appreciated "heads up," allowing management to discuss the facts and proactively devise a dispute resolution strategy.

INFORMAL DISCUSSION

One of the least demanding dispute resolution strategies is informal discussion. That may well be the best and lowest cost strategy for the example above. The parties simply open a discussion about perceived problems and attempt to reach a mutually acceptable resolution. Sometimes informal discussion will show the aggrieved party that a perception was based on a misconception, and the issue is quickly quieted. The main theme of this process is the sense of partnership that a client and engineer should have. It should be solidly established at the outset that the discussion's main purpose is to keep the project on track and that the parties should not be harmed in any way. Personal biases should be left at the doorstep. Discussion should be open and frank but not confrontational. The discussion should be geared toward examination of the alleged facts associated with the issue, eliminating any misconceptions or factual errors and bringing unrecognized facts to the table.

In this process, the costs of resolution are the least, as long as both parties are sincerely interested in reaching agreement and in going forward with the project. It is imperative that the parties are not there just to win an argument for their own personal pride. That attitude seldom leads to a successful resolution, and the dispute can require a higher level of dispute resolution.

NEGOTIATION

Negotiation is a more formalized process than informal discussion, but it is still a less costly procedure than most others. Most guidelines for negotiating point out three primary elements within a successful negotiation process:

- Negotiation is voluntary. It will only succeed if the parties are committed to honest, sincere efforts to reach a negotiated resolution.
- The parties to negotiation must be open to modifying their existing relationship to reduce or eliminate the contentious matter at hand.
- The ultimate goal of negotiation is to settle the dispute in a way that gives each party a situation better than one where negotiation fails.

For negotiation to be effective, the parties must be adequately prepared. Some of the fundamental preparatory elements include the following:

- Each side must have a person authorized to make decisions. If that person is not intimately involved in the project, then someone who is adequately informed should accompany the decision maker.
- Each side should set specific goals. Goals might include the approximate amount of acceptable financial settlement, time extensions, procedures for eliminating future claims over the specific issue at hand, and so on.
- Each side should have some range of potential settlement values.
- Each side should consider elements within the dispute, such as potential financial damages, compilation of facts known to each side, legal constraints, and consequences of failing to reach a settlement, and so forth.
- Each side should size up the other side, if possible, as to personality traits and as to their objectives.
- Each side should consider the extent of documents and exhibits that it should bring to facilitate the process.

Negotiation of disputes has a good record of success in the industry. Well-prepared negotiators generally succeed. Some negotiation scenarios are based on what is known as "positional bargaining," in which the parties start out with demands that are well in excess of their true expectations. This approach is not uncommon in the real estate market, where homes are listed at prices well above that which the seller will gladly accept. That process is not always useful in resolving a dispute. The preferable technique is known as "interest-based bargaining." In that approach, the focus is on the problem, not on the negotiators. The parties explore a variety of solutions and apply objective logic toward selecting a mutually agreeable solution.

STANDING NEUTRAL

In this nonadjudicative process, the parties to a contract for engineering services and/or construction services (more frequently) mutually select an independent third party to resolve conflicts that may arise as a project unfolds. The person selected should be well versed in engineering and construction practices and should have a proven reputation for ethical conduct. This neutral party, or "standing neutral," is given all pertinent contracts and project documents and is often invited to attend project meetings.

Compensation for the standing neutral is normally shared between the parties to the contract. The parties determine whether the standing neutral's dispute resolution opinions will be binding or nonbinding. The decision to establish a standing neutral for a project does, of course, create expense for the parties, and it should be done by written agreement that establishes the guidelines for the standing neutral's role, compensation, and duration of assignment. The standing neutral format provides a mechanism for resolving disputes on an ongoing and rapid basis, thereby mitigating ancillary expenses that can arise because of project delays and legal costs.

DISPUTE REVIEW PANEL

This form of nonadjudicative dispute resolution provides for a panel of knowledgeable experts, usually three in number. The process is particularly well suited to large construction projects. The dispute review board operates in much the same way as a standing neutral. The project owner and the construction contractor normally share costs.

MEDIATION

Mediation is at the high end of the expense range for a nonadjudicative dispute resolution process. Mediation is a procedure whereby a third party who has no specific interest in the project will serve as a facilitator for the parties who wish to resolve a dispute. The mediator's function is to aid the parties to focus on and clarify the areas of concern in the dispute and to prompt the parties to work diligently toward a reasoned resolution.

Mediation can be informal, to the extent that the parties agree to proceed and then select a mutually agreed-upon mediator, who will then set the rules to suit the situation. A more formalized mediation process can be implemented by following recognized rules, such as the American Arbitration Association's (AAA's) rules for mediation.

A mediation proceeding consists of an initial meeting of the mediator with all the parties to the dispute. At the joint meeting, the parties will state their complaints and positions, ordinarily using exhibits, documents, PowerPoint® presentations, photos, and more to support their positions. When that phase is complete, the individual parties go to private areas to discuss their positions. Then the mediator holds confidential meetings with the parties. The mediator strives to aid the parties to focus on issues and eliminate stumbling blocks that can prevent reaching a negotiated settlement. A good mediator needs to have the training, background, and skill to identify "deal breakers" and to effectively encourage the parties to be flexible in eliminating them. The parties must be constantly reminded that should mediation fail, they will be faced with the need to escalate into some form of adjudicative dispute resolution with the concomitant expectation of considerable expense and loss of time.

Although mediation, in the context of this segment, is voluntary, a growing trend is for courts to order mediation before allowing the parties to move forward with full-blown litigation.

ARBITRATION

If all of the nonadjudicative dispute resolution tactics fail, then the parties move into adjudicative dispute resolution, where the parties present their case to a third authoritative party who renders a binding decision. Arbitration is generally recognized by the legal profession as a quicker and less expensive route to a decision, although there is no overwhelming consensus that it is superior to litigation. I have heard attorneys say that arbitrators are prone to "cutting the baby in half," a reference to the biblical story of King Solomon's wisdom in resolving a dispute. Arbitration offers a streamlined version of litigation, with some of these distinctions being:

- Arbitration does not allow the drawing in (joinder) of third parties. In litigation, during the discovery period, it sometimes becomes apparent that a third party, other than the plaintiff or original defendant, may bear some responsibility for the alleged damages, and an amended complaint will be filed to draw in that person (third-party complaint).

- Litigation has a discovery period during which the parties can require each other to answer formal written questions (interrogatories); open their files to inspection; produce copies of documents, files, and records; and appear to give sworn testimony (depositions). None of this is required in arbitration, although the parties may exchange documents.

- Arbitration hearings are less formal than trials. It is not even required that the parties be represented by counsel, although prudent ones commonly are. Rules of evidence are less stringent in arbitration as are the procedures.

- Prejudgment settlements, while common in litigation, are not so in arbitration. A large percentage of arbitration proceedings run through to completion and to decisions.

- Decisions rendered by an arbitrator are final and are not subject to appeal, unlike those of a judge or of a jury in litigation. The only exception to that principle is if it can be shown that an arbitrator was unduly influenced or was guilty of some form of misconduct in reaching the decision. It would also be possible to vacate a verdict if it could be proven that it was reached based upon corruption or fraud.

In a typical arbitration action, the parties can choose their own arbitrator, or they can file a demand for arbitration with the AAA. The AAA will then provide a list of prospective arbitrators from a panel of prequalified experts. The parties have the option to strike names from the list for any reason. Following that, AAA will appoint one of the panelists remaining on the list. AAA will then schedule hearings on dates that work for the parties.

Arbitration hearings resemble litigation, but, as indicated above, the rules of evidence and hearing procedures are less formal than are those in litigation. The arbitrator is required to issue a decision and award within 30 days after the close of the hearings. To give a real-life example of what can happen at the end of a "bench trial" (where finder of fact is a judge rather than a jury), we can look at a case that was concluded in early May 2004 in which I served as an expert. A lot of money was riding on the judge's decision. The judge did not issue his decision until the summer of 2005, leaving the parties in a state of suspense for over a year. In arbitration, that is not permitted.

Arbitration is more costly and time consuming than all of the nonadjudicative procedures discussed above, but the granddaddy of all dispute resolution procedures is litigation.

LITIGATION

Litigation: "A contest in a court of justice for the purpose of enforcing a right or seeking a remedy . . . ."

This definition, found in *Black's Law Dictionary,* clinically describes what can be an exasperating experience for the design professional. Litigation is disruptive, painful, and very expensive. Black's "contest" is closer to a war. The combatants in a construction case may include the owner, engineer, general contractor, subcontractors, and material suppliers. Suits may be filed by one against the other, but often there are multiple defendants and/or cross-claims. Also, third parties outside of the construction loop are increasingly suing one or several of the parties to the project.

Litigation commences when one party feels it has been damaged or injured in some way and can't resolve the matter through any of the dispute resolution techniques discussed above. At that point, assuming that the alleged damage or injury has occurred within the statutory periods established by law, the party files a complaint against the perceived wrongdoer(s) in a court of law. That party is known as a plaintiff. The party or parties against whom the complaint is filed are known as defendant(s). Once the complaint is filed, the defendant(s) are served the complaint, usually by an officer of the court such as a sheriff or constable.

Once served, the defendant(s) have a specified time to file an answer, or, failing that, they can lose the case by default. In the answer, the defendant may counterclaim against the plaintiff. In some jurisdictions, if the complaint is against an engineer or other design professional, the plaintiff must furnish a certificate or affidavit of merit issued by another design professional stating that there is reason to go forward with the litigation. The details of the expert's requirements in preparing such a statement vary from state to state.

After the preliminary filing of complaints, answers, and counterclaims has occurred, the process of discovery begins. Discovery is that process by which litigants are empowered to seek and obtain information relevant to the case. Because the litigants don't always know exactly what is relevant, the rules of discovery are usually quite liberal. Discovery rules can vary from state to state, and there is a set of rules for cases heard in federal courts. Fundamentally, however, discovery allows all parties to the litigation to conduct in-depth investigations in their pursuit of truth. No party is entitled to withhold findings from the others. "Perry Mason" last-ditch discoveries introduced as the jury is about to retire for deliberations are seldom allowed in real life.

Discovery has several main components:

- Interrogatories
- Depositions
- Document demands

Interrogatories

Interrogatories are customarily served early in discovery. The interrogatory is a written compilation of questions, often voluminous and exhaustive, which the parties use to elicit information from each other. This process typically seeks information regarding the names, addresses, and roles of people familiar with the case, and it probes for facts or alleged occurrences upon which the parties will rely. Often interrogatories query what plans, documents, texts, reference materials, and the like were relied upon. They may query ownership of businesses, insurance coverage, business relationships, and more.

Normally, attorneys for the parties serve interrogatories upon the attorneys for other parties. The parties prepare responses with their attorneys. Some questions may be judged by the attorney to be too broad or onerous to answer, or they may be technically improper. Responding to interrogatories is nearly always a joint venture of the attorney and the party. Unlike depositions or courtroom examinations, interrogatories allow time for cool deliberation and reflection in response. Interrogatories are sworn testimony, are admissible in court, and may be used to impeach the credibility of the responder.

Depositions

Depositions usually occur in the later stages of discovery. An attorney issues a subpoena. Subpoenas may, of course, require a court appearance. They are also used to require persons, including expert witnesses, to appear at a specific time and place to be examined under oath. Some subpoenas require that the witness bring certain records to the deposition.

In a deposition proceeding, the witness appears, usually accompanied by counsel. All parties are notified of the deposition, so attorneys for all parties may be present. A court stenographer is also present, and the proceedings are recorded. A transcript is produced for use by any party to the case. The witness is sworn and testifies under oath no less binding than that made on the stand in a courtroom.

The witness is examined first by the attorney who issued the subpoena, but all attorneys representing parties have the opportunity to question the witness as well. While not as rigorously organized as courtroom testimony, there is still a potential for attorneys to squabble, object, or even order the witness not to answer. Even in the absence of such activity, a deposition can be wearing, even exhausting, for the witness. Depositions present a great opportunity for witnesses to contradict their interrogatory responses or even their own statements given during deposition. This is particularly true when a deposition is adjourned and continued at a later time. Unlike the friendly and reassuring environment of the interrogatory response process, depositions can present a hostile and unnerving setting for the witness. Because the deposition transcript can be utilized in court, it can become a significant weapon in challenging subsequent courtroom testimony by the witness.

Document Demands

The demand for documents can be very broad. As an engineer who practices in the forensic field, I can tell you that the amount of paper that is given to me to review in analyzing a case can be overwhelmingly voluminous, sometimes filling many legal storage boxes. Frequently, counsel for the parties will visit the offices of other parties and go through their files and records, marking items for reproduction. Those documents need to be distributed to counsel for all parties, so the volume of paper in a large and complicated matter can be enormous. There are also other sources of documents, such as police reports; OSHA citations; various laws, regulations, codes, and standards; expert reports; supporting pages or chapters from scholarly treatises; and more. If many parties are involved in a litigation matter, all of them may ask to review all of your firm's documents on different occasions. This, of course, can be very disruptive to day-to-day operations, as well as costly in terms of loss of billable time for those who need to participate in the process and in administrative costs. Needless to say, litigation is not only an unpleasant experience but an expensive one as well.

Trial

If a matter proceeds all the way through discovery, it may actually go to trial. Most of the time, however, cases are settled before trial. The parties begin to see the strengths and weaknesses of each other's positions, and expert reports are issued, allowing the parties to see what an independent, objective professional has reviewed, analyzed, and opined upon. Answers to interrogatories, deposition transcripts, and previously unknown documents come to the attention of the parties, perhaps softening their conviction regarding their perceived positions. Furthermore, the trial itself can consume huge chunks of time and money. Attorneys for the parties, as well as insurance carriers and others, have a sense of the impending costs, and, realizing that trial outcomes are almost never predictable to a certainty, they estimate a reasonable settlement scenario as measured against those costs. Settlements are never admissions of wrongdoing, so engineering firms are sometimes motivated to offer settlement in amounts less than the estimated cost of trial.

If the trial proceeds, it can be held before a jury (most commonly), with a judge acting as the referee in legal matters but not participating in the formation of a decision. Sometimes, however, the parties will waive a jury trial in favor of a bench trial, putting the outcome in the hands of a judge.

Once a jury is chosen, the trial opens with the plaintiff's presenting its case. The plaintiff has the burden of proof, meaning that it needs to convince the jury that the defendant caused harm to the plaintiff, and that the harm can be quantified in monetary terms. Following the plaintiff's case, the defendant presents its arguments to the contrary. Both sides present witnesses to support their positions. When the plaintiff presents witnesses, plaintiff's counsel asks questions of the witnesses (direct examination), and defendant's counsel may question those witnesses when their direct testimony is completed (cross-examination). The procedure is repeated during the defendant's case, with the roles of the attorneys reversed.

The rules of evidence in court proceedings are strict and are embodied in rulebooks. The attorneys usually make many challenges with respect to admissibility of evidence, and the judge must puts on a referee hat to decide admissibility. Counsel also objects to questions asked by the opposition with respect to introduction of elements not revealed during discovery and more. I have been on the witness stand several times, sitting silently over prolonged periods, while the attorneys debated positions related to my testimony before the judge. Sometimes the proceedings took on the aura of a bitterly contested debate at which I was merely a spectator.

When all of the allowable testimony and evidence has been exhaustively introduced, the attorneys give their closing remarks to the jury. At that time, they resemble actors playing to a small audience. At the conclusion of closing remarks, the judge gives instructions to the jury regarding the law they are to consider and by which they are to be guided in reaching their verdict. In most jurisdictions, the jury instructions for professional engineering matters include a briefing on the criteria to be considered in a professional negligence matter. The jury then retires to the jury room to deliberate and return with a decision.

If anyone is aggrieved at the findings of the jury, they can appeal the matter to the appeals court. An appeal must be based on an alleged deviation from proper legal conduct by the court. You cannot appeal simply because you lost. The appeal must be based upon procedural errors by the court.

There was only one major lawsuit against the firm I owned for nearly 30 years. That suit resulted in a jury finding in our favor (i.e., there was no negligence and, therefore, no damages to be paid to the plaintiff). The case ran on for a period of nearly seven years, concluding with a trial that spanned just nine days in court. The financial cost for us to prevail was about $250,000 ($25,000 deductible paid by the firm, about $50,000 in billable time by staff during discovery and trial, and about $175,000 in payments by our insurance carrier for legal and ancillary fees).

Based on everything you have seen in this section, it should be obvious that dispute resolution is best managed at an early stage, using one of the least costly nonadjudicative strategies. We don't want this section to cause you to be overly timid about being an engineer. Knowledge is a fantastic weapon to fend off disastrous occurrences. Knowing dispute resolution techniques as well as risk management skills will help you to become an even better engineer than you would if you lacked that knowledge. We want to urge you to keep your eye out for opportunities to learn more on this topic. Seminars are given periodically on risk management and legal obligations. Try to get to one or two in your formative years and continue the educational process periodically throughout your career.

In this book, we have given you a broad overview of the engineering world and guidance that will help you throughout your career. Although some of our advice is not specific to the field of engineering, each subject covered is an essential ingredient to a successful career in engineering. We urge you to take those areas of career guidance seriously and to make them part of the foundation of your career.

Engineering is truly a proud profession. Every person in the civilized world is the beneficiary of what engineers do. Just a simple task like brushing your teeth involves the work of plastics engineers (toothbrush), manufacturing engineers (toothbrush and toothpaste container), civil engineers (piping; valves; and clean, safe water), chemical engineers (toothpaste), and electrical engineers (vibrating or rotating toothbrush, lighting in the bathroom).

Whether you choose to be a civil engineer or an electrical engineer, you should never forget the opportunity and duty you have to contribute to the health, welfare, and protection of the public, and you should never forget that you can and will have a positive influence over the life of many people.

We have urged you to consider licensure as a goal and to participate in your profession through membership in at least one professional association or society. We have urged you to commit yourself to lifetime learning. We have both subscribed to those principals. Our professional careers are, without question, enhanced by those activities. Please keep our encouragements in mind.

While we have presented a number of guidelines for building a successful career, we remind you that engineering is not a static profession. It is dynamic—subject to change and expansion on a steady, seemingly accelerating basis. Some of our information is constant in nature, but you would be well advised to keep your eyes open for change, which means staying on top of technology, continuing education, and professional development.

We hope this book will serve as a faithful companion and a useful resource while you transition into your career and that it will contribute to an expedited professional growth curve. We both feel that our careers would have been greatly enhanced had we known then what we know now about engineering principles. We hope that we have helped you and that you will have a long, successful, and highly enjoyable career in engineering.

May your career design be crafted with the same care and attention that you will place into any product design. It is all in your hands. Good luck.

ENGINEERING ACRONYMS

AAA American Arbitration Association

AAAS American Association for the Advancement of Science

AAEE American Academy of Environmental Engineers

AACE American Association of Cost Engineers

AAES American Association of Engineering Societies

AASHTO American Association of State Highway and Transportation Officials

*ABA American Bar Association

ABET *formerly* Accreditation Board for Engineering and Technology

ACEC American Council of Engineering Companies

ACSM American Congress of Surveyors and Mappers

*AFE Association for Facilities Engineers

AGC Associated General Contractors of America

AIA American Institute of Architects

AIAA American Institute of Aeronautics and Astronautics

AIChE American Institute of Chemical Engineers

AICP American Institute of Certified Planners

AIME American Institute of Mining and Metallurgical Engineers

AIP American Institute of Planners

AIPE American Institute of Plant Engineers

AISES American Indian Science and Engineering Society

ANS American Nuclear Society

ANSI American National Standards Institute

ARTBA American Road and Transportation Building Association

ASAE American Society of Association Executives (American Society of Agricultural Engineers)

* Acronyms used in this book

ASCE American Society of Civil Engineers

ASEE American Society for Engineering Education

ASFE Association of Engineering Firms Practicing in the Geosciences

ASHRAE American Society of Heating, Refrigerating and Air-Conditioning Engineers

ASME American Society of Mechanical Engineers

ASQC American Society for Quality Control

*ASTM American Society for Testing and Materials

*BPO Business Process Offshoring

*CAE Certified Association Executive

CERCLA Comprehensive Environmental Response, Compensation, and Liability Act

CERF Civil Engineering Research Foundation

CESB Council of Engineering Specialty Boards

CESSE Council of Engineering and Scientific Society Executives

CIEP Council for International Engineering Practice

CMAA Construction Management Association of America

COFPAES Council on Federal Procurement of Architectural & Engineering Services

*CPC Continuing Professional Competency

CSI Construction Specifications Institute

*DAB Design-Award-Build

DARPA Defense Advanced Research Projects Agency

*DB Design-Build

EEC Engineering Education Coalition

EIRC Education Information and Resource Center

EJCDC Engineers Joint Contract Documents Committee (includes NSPE, ASCE, and ACEC)

*ELSES Engineering and Land Surveying Services

*ELQTF Engineering Licensure Qualifications Task Force

FAR Federal Audit Regulations

*FE Fundamentals of Engineering test

FHwA Federal Highway Administration

FLSA Fair Labor Standards Act

* Acronyms used in this book

FmHA Farmers Home Administration

GAO General Accounting Office

GSA General Services Administration

ICED Interprofessional Council on Environmental Design

IEEE Institute of Electrical and Electronics Engineers, Inc.

IIE Institute of Industrial Engineers

ISA Instrument Society of America

ISTEA Intermodal Surface Transportation Efficiency Act

ITC International Trade Commission

ITE Institute of Transportation Engineers

JETS Junior Engineering Technical Society

*LQOG Licensure Qualifications Oversight Group

*MCPC Mandatory Continuing Professional Competency

NAAG National Association of Attorneys General

NABIE National Academy of Building Inspection Engineers (referred to as a chartered affinity group)

NACE National Association of Corrosion Engineers

NAE National Academy of Engineering

NAFE National Academy of Forensic Engineers (referred to as a chartered affinity group)

NAS National Academy of Sciences

NASA National Aeronautics and Space Administration

NAVFAC Naval Facilities Engineering Command

NAWIC National Association of Women in Construction

NBS obsolete (National Bureau of Standards); now NIST

NCARB National Council of Architectural Registration Boards

NCEER National Center for Earthquake Engineering Research

NCEES National Council of Examiners for Engineering and Surveying

NCSBCS National Council of States on Building Codes and Standards

NIBS National Institute of Building Sciences

NICE National Institute of Ceramic Engineers

NICET National Institute for Certification in Engineering Technologies

NIEE National Institute for Engineering Ethics (NSPE)

NIEMS National Institute for Engineering Management and Systems (NSPE)

* Acronyms used in this book

NIST National Institute of Standards and Technology (formerly NBS)
NPDES National Pollutant Discharge Elimination System
NPDRE National Professional Development Registry for Engineers
NRC National Research Council or Nuclear Regulatory Commission
NSF National Science Foundation
NSPE National Society of Professional Engineers
OFPP Office of Federal Procurement Policy
OSHA Occupational Safety and Health Administration
*PLI Professional Liability Insurance
*QBS Qualification-Based Selection
*QA/QC Quality Assurance and Quality Control
RCRA Resource Conservation and Recovery Act
*RFP request for proposal
*RFQ request for qualifications
*RPR Resident Project Representative
SAE Society of Automotive Engineers
SME Society of Manufacturing Engineers
SWE Society of Women Engineers
*UPLG Uniform Procedures and Legislative Guidelines
USAAG U.S. Agency Advisory Group.
USCIEP U.S. Council for International Engineering Practice
*VE value engineering
WEF Water Environment Federation (formerly WPCF Water Pollution Control Federation)
WISE Washington Internships for Students in Engineering
*YEAC Young Engineers Advisory Committee

Interest Groups within NSPE

PEC Professional Engineers in Construction
PEE Professional Engineers in Education
PEG Professional Engineers in Government
PEI Professional Engineers in Industry
PEPP Professional Engineers in Private Practice

* Acronyms used in this book

SAMPLE RESUME #1

Frederick P. Norton, PE, PP

Consulting Engineer

3214 West End Avenue, Maple Shade, New Jersey 07123

201-489-5777

EXPERIENCE

February 2003 to present

Self-Employed, CONSULTING ENGINEER CONSULTANT—Established consulting engineering business specializing in residential, commercial, and industrial land development projects, including preparation and design of site plans and subdivisions, design of drainage and detention facilities, septic design, utility layout, grading and soil erosion control design, environmental impact reports, wetland delineations, processing of permits, and presentation of plans to local, state, and federal agencies and boards of approval.

September 1998 to February 2003

Cottonwood Engineering, Cherry Hill, NJ, PROJECT ENGINEER— Designed and managed residential and industrial subdivisions and site developments, including preparation and design of site, grading, utility, and soil erosion and sediment control plans and supporting calculations. Design of retaining walls and detention structures. Environmental impact statements. Expert testimony provided at planning boards, boards of adjustment, and boards of health. Municipal engineering review of projects located in Cherry Hill and Haddonfield, NJ. Construction specification and document preparation. Major projects included:

Sewell Farm, Florence Township, NJ, 64-lot residential subdivision with limited income housing layout and design of storm water management facilities and septic systems, EIS, project management, board of health testimony.

Solomen Associates, Crotin, NJ, 62-acre, 9-lot industrial subdivision—design of storm water management, sanitary sewer and potable water systems, wetlands delineation, EIS, project management, NJDEP permits, expert testimony at planning board, site plan preparation.

Presidential Court, Manahawkin, NJ, 24-townhouse site development on 9 acres, wetlands delineation, NJDEP permits, design of roadways, grading, utilities, EIS, project management, expert testimony at planning board.

May 1995 to September 1998
Hannity Consulting, East Windsor, NJ, PROJECT MANAGER—Designed and administered residential subdivisions, commercial and industrial site developments. Preparation of site, grading, landscaping, soil erosion and sediment control plans, percolation tests, cost estimates, and EIS reports. Road, storm drainage, septic and detention basin design. Testified at public hearings.

July 1991 to May 1995
Reddy Electric Company, El Paso, TX, ENGINEER TRAINEE—CIVIL ENGINEERING DEPARTMENT—Sited and licensed geothermal power plants, including scheduling and cost estimating. Other responsibilities included implementing the state-mandated Compliance Monitoring Program for existing units, ensuring compliance with federal, state, and local laws and ordinances for the construction and operation of a geothermal power plant.

LICENSES
Professional Civil Engineer, Texas
Professional Engineer, New Jersey
Professional Planner, New Jersey

EDUCATION

Drexel University, Philadelphia, PA

B.S. Civil Engineering, June 1991—Studies included surveying, foundations, seepage and earth structures, soil dynamics, earthquake engineering, hydrology, open channel hydraulics, and water resources engineering. GPA: 3.6

CONTINUING EDUCATION

Detention/Retention in Urban Surface Water Management, ASCE, July 2004

Soils of New Jersey, Cook College, September 2000

Environmental Law and Regulation, Cook College, March 1999

On-site Wastewater Disposal Systems—Design, Operation, and Maintenance, Cook College, February 1997

Storm Water Management for Engineers, Cook College, September 1996

PROFESSIONAL ASSOCIATIONS

American Society of Civil Engineers—Member

National Society of Professional Engineers—Member

New Jersey Society of Professional Engineers—Member

HONORARY SOCIETIES

Phi Beta Kappa

Tau Beta Pi

Chi Epsilon

Phi Eta Sigma

AWARDS

Camden County Young Engineer of the Year, 1998

SAMPLE RESUME #2

Allan S. Smith

494 Taylor Hall Eastern State University Sandusky, OH 59321 708-555-1212 *Smith73@esu.edu*	37 Hargrove Avenue Shaker Heights, IL 11654 569-945-4321 *AlanSS@Hotmail.com*

Objective: An internship in manufacturing engineering area

Education: Bachelor of Science in Mechanical Engineering

Eastern State University, Sandusky, OH

Expected date of graduation: June 2007; GPA: 3.6/4.0

Computer Skills: Windows XP, MS Word, Excel, Access, C++

Employment: *June 2006–August 2006*

Production Engineer, Robbins Plastics, Sandusky, OH. Developed new procedures and systems for injection molding machines.

June 2005–August 2005

Machine Operator, Robbins Plastics, Sandusky, OH

Honors/Activities: Dean's List (4 semesters), ESU Athletic Scholarship, ESU baseball team, tennis

References: Available upon request

240.15 Rules of Professional Conduct (NCEES)

A. Licensee's Obligation to Society
  1. Licensees, in the performance of their services for clients, employers, and customers, shall be cognizant that their first and foremost responsibility is to the public welfare.
  2. Licensees shall approve and seal only those design documents and surveys that conform to accepted engineering and surveying standards and safeguard the life, health, property, and welfare of the public.
  3. Licensees shall notify their employer or client and such other authority as may be appropriate when their professional judgment is overruled under circumstances where the life, health, property, or welfare of the public is endangered.
  4. Licensees shall be objective and truthful in professional reports, statements, or testimony. They shall include all relevant and pertinent information in such reports, statements, or testimony.
  5. Licensees shall express a professional opinion publicly only when it is founded upon an adequate knowledge of the facts and a competent evaluation of the subject matter.
  6. Licensees shall issue no statements, criticisms, or arguments on technical matters which are inspired or paid for by interested parties, unless they explicitly identify the interested parties on whose behalf they are speaking and reveal any interest they have in the matters.
  7. Licensees shall not permit the use of their name or firm name by, nor associate in the business ventures with, any person or firm which is engaging in fraudulent or dishonest business or professional practices.
  8. Licensees having knowledge of possible violations of any of these Rules of Professional Conduct shall provide the board with the information and assistance necessary to make the final determination of such violation.
B. Licensee's Obligation to Employer and Clients
  1. Licensees shall undertake assignments only when qualified by education or experience in the specific technical fields of engineering or surveying involved.
  2. Licensees shall not affix their signatures or seals to any plans or documents dealing with subject matter in which they lack competence, nor to any such plan or document not prepared under their direct control and personal supervision.
  3. Licensees may accept assignments for coordination of an entire project, provided that each design segment is signed and sealed by the licensee responsible for preparation of that design segment.
  4. Licensees shall not reveal facts, data, or information obtained in a professional capacity without the prior consent of the client or employer except as authorized or required by law. Licensees shall not solicit or accept gratuities, directly or indirectly, from contractors, their agents, or other parties in connection with work for employers or clients

5. Licensees shall make full prior disclosures to their employers or clients of potential conflicts of interest or other circumstances which could influence or appear to influence their judgment or the quality of their service.

6. Licensees shall not accept compensation, financial or otherwise, from more than one party for services pertaining to the same project, unless the circumstances are fully disclosed and agreed to by all interested parties.

7. Licensees shall not solicit or accept a professional contract from a governmental body on which a principal or officer of their organization serves as a member. Conversely, licensees serving as members, advisors, or employees of a government body or department, who are the principals or employees of a private concern, shall not participate in decisions with respect to professional services offered or provided by said concern to the governmental body which they serve.

C. Licensee's Obligation to Other Licensees

1. Licensees shall not falsify or permit misrepresentation of their, or their associates', academic or professional qualifications. They shall not misrepresent or exaggerate their degree of responsibility in prior assignments nor the complexity of said assignments. Presentations incident to the solicitation of employment or business shall not misrepresent pertinent facts concerning employers, employees, associates, joint ventures, or past accomplishments.

2. Licensees shall not offer, give, solicit, or receive, either directly or indirectly, any commission, or gift, or other valuable consideration in order to secure work, and shall not make any political contribution with the intent to influence the award of a contract by public authority.

Licensees shall not attempt to injure, maliciously or falsely, directly or indirectly, the professional reputation, prospects, practice, or employment of other licensees, nor indiscriminately criticize other licensees' work.

NSPE CODE OF ETHICS FOR ENGINEERS
(Effective Jan. 2006)

Preamble

Engineering is an important and learned profession. As members of this profession, engineers are expected to exhibit the highest standards of honesty and integrity. Engineering has a direct and vital impact on the quality of life for all people. Accordingly, the services provided by engineers require honesty, impartiality, fairness, and equity, and must be dedicated to the protection of the public health, safety, and welfare. Engineers must perform under a standard of professional behavior that requires adherence to the highest principles of ethical conduct.

I. Fundamental Canons

Engineers, in the fulfillment of their professional duties, shall:

1. Hold paramount the safety, health, and welfare of the public.
2. Perform services only in areas of their competence.
3. Issue public statements only in an objective and truthful manner.
4. Act for each employer or client as faithful agents or trustees.
5. Avoid deceptive acts.
6. Conduct themselves honorably, responsibly, ethically, and lawfully so as to enhance the honor, reputation, and usefulness of the profession.

II. Rules of Practice

1. Engineers shall hold paramount the safety, health, and welfare of the public.

   a. If engineers' judgment is overruled under circumstances that endanger life or property, they shall notify their employer or client and such other authority as may be appropriate.

   b. Engineers shall approve only those engineering documents that are in conformity with applicable standards.

   c. Engineers shall not reveal facts, data, or information without the prior consent of the client or employer except as authorized or required by law or this Code.

   d. Engineers shall not permit the use of their name or associate in business ventures with any person or firm that they believe is engaged in fraudulent or dishonest enterprise.

   e. Engineers shall not aid or abet the unlawful practice of engineering by a person or firm.

   f. Engineers having knowledge of any alleged violation of this Code shall report thereon to appropriate professional bodies and, when relevant, also to public authorities, and cooperate with the proper authorities in furnishing such information or assistance as may be required.

2. Engineers shall perform services only in the areas of their competence. Engineers shall undertake assignments only when qualified by education or experience in the specific technical fields involved.

   a. Engineers shall not affix their signatures to any plans or documents dealing with subject matter in which they lack competence, nor to any plan or document not prepared under their direction and control.

   b. Engineers may accept assignments and assume responsibility for coordination of an entire project and sign and seal the engineering documents for the entire project, provided that each technical segment is signed and sealed only by the qualified engineers who prepared the segment.

3. Engineers shall issue public statements only in an objective and truthful manner.

   a. Engineers shall be objective and truthful in professional reports, statements, or testimony. They shall include all relevant and pertinent information in such reports, statements, or testimony, which should bear the date indicating when it was current.

b. Engineers may express publicly technical opinions that are founded upon knowledge of the facts and competence in the subject matter.

c. Engineers shall issue no statements, criticisms, or arguments on technical matters that are inspired or paid for by interested parties, unless they have prefaced their comments by explicitly identifying the interested parties on whose behalf they are speaking, and by revealing the existence of any interest the engineers may have in the matters.

4. Engineers shall act for each employer or client as faithful agents or trustees.

a. Engineers shall disclose all known or potential conflicts of interest that could influence or appear to influence their judgment or the quality of their services.

b. Engineers shall not accept compensation, financial or otherwise, from more than one party for services on the same project, or for services pertaining to the same project, unless the circumstances are fully disclosed and agreed to by all interested parties.

c. Engineers shall not solicit or accept financial or other valuable consideration, directly or indirectly, from outside agents in connection with the work for which they are responsible.

d. Engineers in public service as members, advisors, or employees of a governmental or quasi-governmental body or department shall not participate in decisions with respect to services solicited or provided by them or their organizations in private or public engineering practice.

e. Engineers shall not solicit or accept a contract from a governmental body on which a principal or officer of their organization serves as a member.

5. Engineers shall avoid deceptive acts.

a. Engineers shall not falsify their qualifications or permit misrepresentation of their or their associates' qualifications. They shall not misrepresent or exaggerate their responsibility in or for the subject matter of prior assignments. Brochures or other presentations incident to the solicitation of employment shall not misrepresent pertinent facts concerning employers, employees, associates, joint venturers, or past accomplishments.

b. Engineers shall not offer, give, solicit, or receive, either directly or indirectly, any contribution to influence the award of a contract by public authority, or which may be reasonably construed by the public as having the effect or intent of influencing the awarding of a contract. They shall not offer any gift or other valuable consideration in order to secure work. They shall not pay a commission, percentage, or brokerage fee in order to secure work, except to a bona fide employee or bona fide established commercial or marketing agencies retained by them.

III. Professional Obligations

1. Engineers shall be guided in all their relations by the highest standards of honesty and integrity.

   a. Engineers shall acknowledge their errors and shall not distort or alter the facts.

   b. Engineers shall advise their clients or employers when they believe a project will not be successful.

   c. Engineers shall not accept outside employment to the detriment of their regular work or interest. Before accepting any outside engineering employment, they will notify their employers.

   d. Engineers shall not attempt to attract an engineer from another employer by false or misleading pretenses.

   e. Engineers shall not promote their own interest at the expense of the dignity and integrity of the profession.

2. Engineers shall at all times strive to serve the public interest.

   a. Engineers shall seek opportunities to participate in civic affairs; career guidance for youths; and work for the advancement of the safety, health, and well-being of their community.

   b. Engineers shall not complete, sign, or seal plans and/or specifications that are not in conformity with applicable engineering standards. If the client or employer insists on such unprofessional conduct, they shall notify the proper authorities and withdraw from further service on the project.

   c. Engineers shall endeavor to extend public knowledge and appreciation of engineering and its achievements.

   d. Engineers shall strive to adhere to the principles of sustainable development[1] in order to protect the environment for future generations.

3. Engineers shall avoid all conduct or practice that deceives the public.

   a. Engineers shall avoid the use of statements containing a material misrepresentation of fact or omitting a material fact.

   b. Consistent with the foregoing, engineers may advertise for recruitment of personnel.

   c. Consistent with the foregoing, engineers may prepare articles for the lay or technical press, but such articles shall not imply credit to the author for work performed by others.

4. Engineers shall not disclose, without consent, confidential information concerning the business affairs or technical processes of any present or former client or employer, or public body on which they serve.

   a. Engineers shall not, without the consent of all interested parties, promote or arrange for new employment or practice in connection with a specific project for which the engineer has gained particular and specialized knowledge.

b. Engineers shall not, without the consent of all interested parties, participate in or represent an adversary interest in connection with a specific projec- or proceeding in which the engineer has gained particular specialized knowledge on behalf of a former client or employer.

5. Engineers shall not be influenced in their professional duties by conflicting interests.

a. Engineers shall not accept financial or other considerations, including free engineering designs, from material or equipment suppliers for specifying their product.

b. Engineers shall not accept commissions or allowances, directly or indirectly, from contractors or other parties dealing with clients or employers of the engineer in connection with work for which the engineer is responsible.

6. Engineers shall not attempt to obtain employment or advancement or professional engagements by untruthfully criticizing other engineers, or by other improper or questionable methods.

a. Engineers shall not request, propose, or accept a commission on a contingent basis under circumstances in which their judgment may be compromised.

a. Engineers in salaried positions shall accept part-time engineering work only to the extent consistent with policies of the employer and in accordance with ethical considerations.

a. Engineers shall not, without consent, use equipment, supplies, laboratory, or office facilities of an employer to carry on outside private practice.

7. Engineers shall not attempt to injure, maliciously or falsely, directly or indirectly, the professional reputation, prospects, practice, or employment of other engineers. Engineers who believe others are guilty of unethical or illegal practice shall present such information to the proper authority for action.

a. Engineers in private practice shall not review the work of another engineer for the same client, except with the knowledge of such engineer, or unless the connection of such engineer with the work has been terminated.

b. Engineers in governmental, industrial, or educational employ are entitled to review and evaluate the work of other engineers when so required by their employment duties.

c. Engineers in sales or industrial employ are entitled to make engineering comparisons of represented products with products of other suppliers.

8. Engineers shall accept personal responsibility for their professional activities, provided, however, that engineers may seek indemnification for services arising out of their practice for other than gross negligence, where the engineer's interests cannot otherwise be protected.

a. Engineers shall conform with state registration laws in the practice of engineering.

b. Engineers shall not use association with a nonengineer, a corporation, or partnership as a "cloak" for unethical acts.
9. Engineers shall give credit for engineering work to those to whom credit is due, and will recognize the proprietary interests of others.
  a. Engineers shall, whenever possible, name the person or persons who may be individually responsible for designs, inventions, writings, or other accomplishments.
  b. Engineers using designs supplied by a client recognize that the designs remain the property of the client and may not be duplicated by the engineer for others without express permission.
  c. Engineers, before undertaking work for others in connection with which the engineer may make improvements, plans, designs, inventions, or other records that may justify copyrights or patents, should enter into a positive agreement regarding ownership.
  d. Engineers' designs, data, records, and notes referring exclusively to an employer's work are the employer's property. The employer should indemnify the engineer for use of the information for any purpose other than the original purpose.
  e. Engineers shall continue their professional development throughout their careers and should keep current in their specialty fields by engaging in professional practice, participating in continuing education courses, reading in the technical literature, and attending professional meetings and seminars.

Footnote 1 "Sustainable development" is the challenge of meeting human needs for natural resources, industrial products, energy, food, transportation, shelter, and effective waste management while conserving and protecting environmental quality and the natural resource base essential for future development.

—*As Revised January 2006*

"By order of the United States District Court for the District of Columbia, former Section 11(c) of the NSPE Code of Ethics prohibiting competitive bidding, and all policy statements, opinions, rulings or other guidelines interpreting its scope, have been rescinded as unlawfully interfering with the legal right of engineers, protected under the antitrust laws, to provide price information to prospective clients; accordingly, nothing contained in the NSPE Code of Ethics, policy statements, opinions, rulings or other guidelines prohibits the submission of price quotations or competitive bids for engineering services at any time or in any amount."

Statement by NSPE Executive Committee

In order to correct misunderstandings which have been indicated in some instances since the issuance of the Supreme Court decision and the entry of the Final Judgment, it is noted that in its decision of April 25, 1978, the Supreme Court of the United States declared: "The Sherman Act does not require competitive bidding."

It is further noted that as made clear in the Supreme Court decision:
1. Engineers and firms may individually refuse to bid for engineering services.
2. Clients are not required to seek bids for engineering services.
3. Federal, state, and local laws governing procedures to procure engineering services are not affected, and remain in full force and effect.
4. State societies and local chapters are free to actively and aggressively seek legislation for professional selection and negotiation procedures by public agencies.
5. State registration board rules of professional conduct, including rules prohibiting competitive bidding for engineering services, are not affected and remain in full force and effect. State registration boards with authority to adopt rules of professional conduct may adopt rules governing procedures to obtain engineering services.
6. As noted by the Supreme Court, "nothing in the judgment prevents NSPE and its members from attempting to influence governmental action . . ."

NOTE: In regard to the question of application of the Code to corporations vis-à-vis real persons, business form or type should not negate nor influence conformance of individuals to the Code. The Code deals with professional services, which services must be performed by real persons. Real persons in turn establish and implement policies within business structures. The Code is clearly written to apply to the Engineer, and it is incumbent on members of NSPE to endeavor to live up to its provisions. This applies to all pertinent sections of the Code.

Douglas E. Benner, P.E., FNSPE

Douglas E. Benner is a licensed Professional Engineer in New Jersey and Georgia. He holds an associate degree in electrical technology from Spring Garden Institute, a bachelor of science in industrial engineering degree from Lehigh University and an MBA from Seton Hall University.

Mr. Benner is currently President of DEB Consulting, which performs electrical engineering and project management services. He has 34 years of varied experience with a major electric utility and has performed project management in plant construction, telecommunications facility construction, electric power system control computers, and technology transfer.

Mr. Benner has served as President of the New Jersey Society of Professional Engineers, and as NJSPE's National Director. He has served on numerous state committees and chaired the task force that created and operates the NJSPE Institute for Professional Lead-

ership. He chaired the New Jersey chapter of professional engineers in industry practice division for six years, and was NJ PEI Governor (national representative) for four years. At NSPE, he has chaired the Constitution and By-laws Committee, the Resource Management Committee, the task force on employment issues, and served on the career transition task force. He also served as vice president for professional engineers in industry.

Mr. Benner has received the following awards and honors:

- New Jersey SPE Distinguished Service Award (1994)
- NSPE Distinguished Member Award (1995)
- NSPE PEI Outstanding Service Award (2003)
- Fellow Member of NSPE (2000)

Mr. Benner resides in Marietta, Georgia with his wife Fabiola. He has four daughters and two granddaughters.

Bernard R. Berson PE, LS, PP, FNSPE

Bernard R. Berson is a licensed Professional Engineer and Land Surveyor in New Jersey, Connecticut (retired status), New York (inactive status), and Pennsylvania. He is a licensed Professional Engineer in Delaware and Massachusetts, and a licensed Land Surveyor in Maryland and Virginia.

He holds a BSCE from the University of Connecticut and an MSCE from Newark College of Engineering. He has practiced in the fields of Engineering and Land Surveying for nearly 50 years, including four years as a commissioned officer in the U. S. Coast & Geodetic Survey. He owned and operated a civil engineering and land surveying practice for nearly 30 years. In February 2000 he ended a long career of full-time employment, and is operating as a consultant specializing in forensic engineering/surveying.

He has served as President of the Association of Practicing Civil Engineers and Land Surveyors, as president of the New Jersey Society of Professional Engineers, and as NJSPE's National Director for three years. He has chaired a number of NJSPE committees, including the Tort Reform Committee and the NJSPE 75th Anniversary task force in 1999. He was a member of the task force that created and operates the NJSPE Institute for Professional Leadership, and was also chairman of the professional con-

duct committee. He chaired the New Jersey chapter of professional engineers in private practice for two years, and was NJ PEPP Governor (national representative) for four years.

At NSPE, he has served on the communications committee, for ten years on the PEPP professional liability committee, and as vice president for private practice. He served for two years as chairman of the NSPE mentoring task force, and as a member of the leadership development task force. He is serving as president-elect of NSPE, and will be installed as president in July 2007.

Mr. Berson has received a number of awards and honors, including:

- U.S. Department of Commerce Distinguished Service Award (1961)
- New Jersey SPE Engineer of the Year (1987)
- NSPE/PEPP Award for Outstanding Contribution to the Advancement and Recognition of Private Practice (1994)
- New Jersey SPE Distinguished Engineer (1995)
- NSPE Distinguished Service Award (1996)
- University of Connecticut School of Engineering Distinguished Alumnus (1997)
- Fellow Member of NSPE (2000)

Mr. Berson resides in Perrineville, New Jersey, with his wife Toby. The couple has four daughters and six grandchildren.